ROBERT JOHNSON
*Monroe Community College,
Rochester, New York*

Study Guide with Self-Correcting Exercises for

Elementary Statistics

Fifth Edition

PWS-KENT Publishing Company
Boston, Massachusetts

Study Guide with Self-Correcting Exercises for

Elementary
Statistics

PWS-KENT
Publishing Company

20 Park Plaza
Boston, Massachusetts 02116

Copyright © 1988 by PWS-KENT Publishing Company.

All rights reserved. No part of this book may be reproduced or transmitted in any form or by any means, electronic or mechanical, including photocopying, recording, or by any information storage and retrieval system, without permission, in writing, from the publisher.

PWS-KENT Publishing Company is a division of Wadsworth, Inc.

Printed in the United States of America

88 89 90 91 92 - 10 9 8 7 6 5 4 3 2 1

ISBN 0-534-91774-7

CONTENTS

Preface	xi	
Introduction	SOME BASIC MATHEMATICAL IDEAS	1
I-1:	The Language of Algebra	1
I-2:	The Use of Formulas	3
I-3:	Random Numbers	7
I-4:	The Coordinate-Axis System and the Equation of a Straight Line	9
I-5:	Tree Diagrams	15
I-6:	Sets and Subsets	19
I-7:	Venn Diagrams	22
I-8:	The Use of Factorial Notation	24
I-9:	The Binomial Coefficient $\binom{n}{r}$	25

chapter 1	STATISTICS	28
1-1:	Basic Terms	28
1-2:	Types of Variables	30

chapter 2 DESCRIPTIVE ANALYSIS AND PRESENTATION OF SINGLE-VARIABLE DATA   34

2-1:	The Five "Averages"	34
2-2:	The Standard Deviation of a Sample	35
2-3:	Locating the Median, the Quartiles, and the Percentiles	39
2-4:	Frequency Distributions	42
2-5:	Calculations Using a Frequency Distribution	44
2-6:	Chebyshev's Theorem	47
2-7:	The Empirical Rule	51

chapter 3 DESCRIPTIVE ANALYSIS AND PRESENTATION OF BIVARIATE DATA   58

3-1:	The Correlation Coefficient	58
3-2:	The Line of Best Fit	60
3-3:	Making Predictions	61

chapter 4 PROBABILITY   67

4-1:	Mutually Exclusive and All-Inclusive Events	67
4-2:	Probabilities of Compound Events Using Relative Frequency	69
4-3:	Independent Events	72
4-4:	The Multiplication Rule	75
4-5:	The Addition Rule	78

chapter 5 PROBABILITY DISTRIBUTIONS (DISCRETE VARIABLES) 86

5-1:	The Mean and Standard Deviation of a Discrete Probability Distribution	86
5-2:	Binomial Probabilities	89

chapter 6 THE NORMAL PROBABILITY DISTRIBUTION 96

6-1:	The Standard Score	96
6-2:	Probabilities Associated with the Normal Distribution	98

chapter 7 SAMPLE VARIABILITY 104

7-1:	The Central Limit Theorem, Theoretically	104
7-2:	The Central Limit Theorem, Empirically	107
7-3:	Application of the Central Limit Theorem	110

chapter 8 INTRODUCTION TO STATISTICAL INFERENCES 115

8-1:	The Nature of Hypothesis Testing	115
8-2:	Alpha and Beta	118
8-3:	The Nature of Estimation	120

chapter 9 INFERENCES INVOLVING ONE POPULATION — 127

- 9-1: Inferences Concerning One Mean — 127
- 9-2: Inferences Concerning One Proportion — 129
- 9-3: Inferences Concerning One Standard Deviation — 131

chapter 10 INFERENCES INVOLVING TWO POPULATIONS — 138

- 10-1: Inferences Concerning the Difference Between Two Independent Means — 138
- 10-2: Inferences Concerning Two Variances — 139
- 10-3: Inferences Concerning the Difference Between Two Independent Means — 140
- 10-4: Inferences Concerning the Difference Between Two Dependent Means — 142
- 10-5: Inferences Concerning the Difference Between Two Proportions — 143

chapter 11 ADDITIONAL APPLICATIONS OF CHI-SQUARE — 150

- 11-1: The Multinomial Experiment — 150
- 11-2: The Contingency Table — 152

chapter 12 ANALYSIS OF VARIANCE — 159

- 12-1: The ANOVA Concept — 159

CONTENTS

chapter 13 LINEAR CORRELATION AND REGRESSION ANALYSIS — 167

- 13-1: The Random Error ϵ — 167
- 13-2: Variance About the Line of Best Fit — 168
- 13-3: Confidence-Interval Estimates for Regression — 170

chapter 14 ELEMENTS OF NONPARAMETRIC STATISTICS — 178

- 14-1: The Concept of Nonparametrics, Demonstrated with the Sign Test — 178
- 14-2: The Mann-Whitney U Test — 181
- 14-3: The Runs Test — 184
- 14-4: Spearman Rank Correlation Coefficient — 187

CONTENTS

Solutions 193

Introduction 193
Chapter 1 203
Chapter 2 205
Chapter 3 233
Chapter 4 242
Chapter 5 251
Chapter 6 259
Chapter 7 267
Chapter 8 276
Chapter 9 282
Chapter 10 289
Chapter 11 299
Chapter 12 307
Chapter 13 318
Chapter 14 335

PREFACE

This study guide is designed to be used with the text *Elementary Statistics*, fifth edition by Robert R. Johnson. Its purpose is to provide a microscopic view of individual concepts for which you may have questions or could use a second explanation. The guide is not intended to be self-contained. The guide is intended to blend with the textbook so that together they will be a more effective teaching device than either used separately. In many cases the lessons in this study guide answer the kinds of questions that you might ask after hearing a lecture or after attempting some of the homework exercises from the textbook.

As you will see, the guide is composed of lessons. Our definition of a lesson is a set of information consisting of an explanation, illustrations, and additional exercises, all of which you should be able to read, comprehend, and answer in a period of fifteen to forty minutes, depending on the length of the lesson. Each lesson deals with a single concept and zeroes in on just the main points. Although relationships between concepts are discussed when appropriate, the major emphasis is on individual concepts.

The lessons are organized into chapters. Each chapter corresponds to the respective chapter in the textbook, except for the Introduction. The Introduction contains ten lessons on topics that one would usually consider background for a statistics course. However, we all need a refresher on some of these topics occasionally.

Notes: 1. Page and chapter references in this study guide refer to the textbook, as do the formula identification numbers.
2. Complete solutions to all exercises in this study guide are included in the back of the guide.

When using this guide, you should work under a general procedure that is at least similar to that outlined here. First, read the textbook. Second, attend lectures and take good notes. Third, reread the text material and study your lecture notes. Fourth, practice these concepts by doing your homework. And fifth, use the vocabulary list and the quiz at the end of each chapter in the textbook to evaluate your own progress. Don't be satisfied just to get the right answer, but know why it is the right one.

PREFACE

Use the study guide to gain insight into concepts that you had difficulty with. The guide is an instructional supplement to go with the text, not to replace it. The study guide may be used specifically as follows: Study the Introduction to the guide sometime before you begin Chapter 2 of the textbook. Then read the other chapters in the guide as the lessons become appropriate. For example, after you study the concept of standard deviation in Chapter 2 of the textbook and attempt the first few exercises involving this concept, turn to the study guide and read a second explanation. Complete the exercises *in the guide* pertaining to standard deviation, then return to the text exercises.

As an additional aid, the guide contains a list of self-correcting exercises in each chapter. By using these exercises, you can test your mastery of the ideas presented in the chapter and also get some additional practice in problem solving. One way to use these exercises might be as follows: When you have completed a chapter, work the first few self-correcting exercises. If you have no difficulty with these, skip down to the middle of the list and work a few more problems. Anytime you find you cannot solve the problems, return to the explanations and discussions in the textbook or the guide. Continue in this way until you can do the exercises toward the end of the list with ease. At that point, you will know you have a good mastery of the ideas presented in that chapter.

When used in the ways outlined above, this guide and its self-correcting exercises will help you to master elementary statistics.

R.R.J.

SOME BASIC MATHEMATICAL IDEAS

INTRODUCTION

LESSON I-1: THE LANGUAGE OF ALGEBRA

Algebra is a language in which letters, numbers, and operational signs replace words in the role of communication. Several terms are used to describe the many aspects of this system of communication. These will be pointed out as we survey the basic concepts of algebraic representation in this lesson.

One of the main points that you want to keep in mind is that letters in algebraic notation represent unknown or variable numbers. For example, x might be used to represent the age of a person. The value of x will change from person to person — thus it is variable. Once an exact situation (person, in this case) is identified, then the value of x may be found and thus becomes a known quantity. Other algebraic expressions can also be used to represent numerical values. $x - 5$ represents some number; in particular, it is that number which is exactly 5 less than whatever the value of x is. $x - 5$ might be interpreted as being a person's age five years ago, but it is still a numerical value.

When discussing algebraic notation, words like product, factor, coefficient, and multinomial are frequently used. What do these terms mean? A *product* is the result of multiplying two or more numbers together. Each number used in obtaining the product is known as a *factor*. The product of the two numbers x and y is written xy, and each of the numbers x and y is a factor of xy. Another example of a product is $5x^2y$, obtained by multiplying 5, x, x, and y together. When one of the factors is a constant (the 5 in this case), this factor is always written first and is called the *coefficient*. If we were to add our first product to our second product $(5x^2y + xy)$ we would have a multinomial expression. A *multinomial* is simply a sum of two or more terms, where each term is a product of numbers.

There are many skills of manipulation that can be learned, although most of them are not necessary for success in business statistics. Basically we only need to be able to read a set of "directions" expressed algebraically.

Suppose that we were given the expressions $2(x - 5)$ and $2x - 5$. Are they the same number? If the value of x is 10, how can we find the value of the numbers $2(x - 5)$ and $2x - 5$? $2(x - 5)$ tells us to subtract 5 from x and then multiply by

2, whereas $2x - 5$ tells us to multiply x by 2 and then subtract 5 from the result. If x is 10, what are these two new numbers? $2(x - 5)$ is 2 times $(10 - 5)$, or 2 times (5), which is 10 [i.e., when $x = 10$, $2(x - 5) = 10$]. However, $2x - 5$ is (2 times 10) - 5, or 20 - 5, which is 15 (i.e., when $x = 10$, $2x - 5 = 15$). As you can see, the two resulting numbers are not the same. These two expressions illustrate how different an answer can be if you misinterpret the meaning of the algebraic expression.

Several of the exercises for this lesson will have you practice (1) reading algebraic expressions and (2) writing algebraic expressions when the verbal translations are given. This very basic representation may well be the single most important fundamental concept. You must be able to read algebraic expressions correctly.

Illustration   Translate $5x - 7$ into words.

Solution   Multiply the number x by 5 and then subtract 7 from that product.

Illustration   Translate $x^2 f$ and $(xf)^2$ into words.

Solution $x^2 f$ means to multiply x by itself and then multiply by f, whereas $(xf)^2$ means to multiply x by f and then multiply this product by itself to obtain the answer.

Illustration   Express this verbal description as an algebraic expression: "Square the sum of x and y."

Solution $(x + y)$ is the sum of x and y; now square it — $(x + y)^2$.

EXERCISES I-1

1. Let y be some number and then express another number which is:
   a. the double of y;
   b. the square of y;
   c. 6 more than the triple of y;
   d. 2 less than the square of y;
   e. 2 more than the square root of y.

2. Consider the following list of four different algebraic expressions and then answer each of the questions.
   I. $2x + 3$
   II. $3xy$
   III. $x^2 f$
   IV. $n(n - 1)$
   a. Which expressions represent the product of numbers?
   b. How many terms are there in expressions I, II, and III?
   c. What is the coefficient in II?
   d. What are the factors for expressions II and IV?

3. Give a verbal translation for each of the following algebraic expressions.
   a. $5(x + 2)$
   b. $5x + 2$
   c. $x^2 + y^2$
   d. $(x + y)^2$

LESSON I-2: THE USE OF FORMULAS

Your ability to use formulas is going to play a major role in your study of statistics. What is a formula? A formula is an algebraic expression that describes the procedure that must be followed in order to find a desired numerical piece of information. The purpose of this lesson is to give you some experience in reading and following the directions set forth by specific formulas.

Basically you can think of a formula as a "compactor machine." We will put certain numerical information into our machine; it will perform the necessary operations and then give us back the desired numerical answer. This little compactor machine does much the same as the formula will do for you when it is followed accurately. It will take the given information and sort it, squeeze it, and do whatever is necessary to it in order to yield the answer.

Below are a few simple formulas. The first shows one piece of input and one resulting piece of output.

Illustration   Formula $i = (n + 1)/2$ (i is equal to n plus 1 divided by 2.) Find i when $n = 10$.

Solution $i = (10 + 1)/2 = 11/2 = \underline{5.5}$

The next illustration shows two input numbers that will yield one answer.

Illustration   Formula: $R = H - L$ (R is equal to the difference of H and L.) Find R when $H = 125$ and $L = 94$.

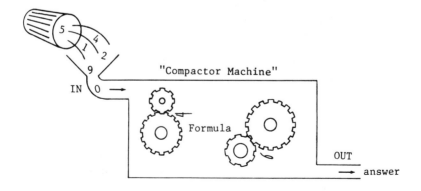

Figure I-1

Solution $R = 125 - 94 = \underline{31}$

EXERCISES I-2

1. Given that $i = (n + 1)/2$, find the value of i in each of the following cases.
   a. $n = 5$
   b. $n = 20$
   c. $n = 100$

2. Given that $R = H - L$, find the value of R in each of the following cases.
   a. $H = 100, L = 50$
   b. $H = 50, L = 14$

3. Given that $z = (a - b)/c$, find the value of z for each of the following cases.
   a. $a = 100, b = 70, c = 10$
   b. $a = 50, b = 60, c = 2$
   c. $a = 19, b = 12, c = 3$

Both of the above illustrations are examples of formulas that require only the use of substitution. That is, we replace the variables (letters) in the formula with a specified value and we evaluate — the result is relatively easy to obtain.

Many statistical formulas use the capital Greek letter sigma (Σ). This symbol stands for the direction, "Sum up the appropriate values." The use of Σ requires an index system as well as the identification of each addend to be used in obtaining the sum. The symbol $\sum_{i=1}^{10} x_i$ means, "The sum of all the values of x assigned to cases one through ten." The index system is the i values. You might think of an i value as the identification number of the source from which each of the x values will be obtained. Let x be the number of seats in a classroom. The x_1 is the number of seats in room one, x_2 is the number of seats in room two, and so on, and the symbol $\sum_{i=1}^{10} x_i$ stands for the total number of seats in rooms one through ten.

Illustration   Consider the following inventory about the number of children per family.

i = Family number	1	2	3	4	5	6	7	8	9	10
x = Number of Children	2	3	2	4	5	6	4	3	3	2

Find the $\sum_{i=1}^{10} x_i$.

USE OF FORMULAS Lesson I-2

Solution

$$\sum_{i=1}^{10} x_i = x_1 + x_2 + x_3 + x_4 \ldots + x_{10}$$

$$= 2 + 3 + 2 + 4 + 5 + 6 + 4 + 3 + 3 + 2 = \underline{34}$$

Find the $\sum_{i=4}^{8} x_i$.

Solution

$$\sum_{i=4}^{8} x_i = x_4 + x_5 + x_6 + x_7 + x_8 = 4 + 5 + 6 + 4 + 3 = \underline{22}$$

Notice that the indexing system identifies the first and last source and you are expected to use those sources and all the sources in between.

This index system must be used whenever only part of the available information is to be used. In statistics, however, we will usually use all of the available information, and to simplify the formulas we will make an adjustment. This adjustment is actually an agreement that allows us to do away with the index system in situations where all values are used. Thus in our previous illustration, the $\sum_{i=1}^{10} x_i$ could have been written simply as Σx.

Note: We must use the index system when only part of the available data is to be used. The lack of the index indicates that all data are being used.

Illustration Given the following six values for x: 1, 3, 7, 2, 4, 5, find the Σx.

Solution $\Sigma x = 1 + 3 + 7 + 2 + 4 + 5 = \underline{22}$

Throughout the study and use of statistics you will find many formulas which use the Σ symbol. The terms in the formulas can be easily misread when terms like Σx^2 and $(\Sigma x)^2$ occur. These two symbols are quite different. Σx^2 [$\Sigma(x^2)$] means square each x and add up the squares. Using the values of x given in the previous illustration, $\Sigma x^2 = 1^2 + 3^2 + 7^2 + 2^2 + 4^2 + 5^2 = 1 + 9 + 49 + 4 + 16 + 25 = \underline{104}$. However $(\Sigma x)^2$ means that the sum of the x's is squared; that is, $(\Sigma x)^2 = (1 + 3 + 7 + 2 + 4 + 5)^2 = (22)^2 = \underline{484}$. As you can see, there is quite a difference between Σx^2 and $(\Sigma x)^2$.

Likewise, the Σxy and the $\Sigma x \Sigma y$ are different. These forms will appear only when there are paired data, as shown in the following illustration.

Illustration Given the following five pairs of data, find Σxy and $\Sigma x \Sigma y$.

x	1	6	9	3	4
y	7	8	2	5	10

Solution $\Sigma x \Sigma y$ means the Σx multiplied by the Σy. The $\Sigma x = 1 + 6 + 9 + 3 + 4 = 23$ and $\Sigma y = 7 + 8 + 2 + 5 + 10 = 32$. Therefore $\Sigma x \Sigma y = (23)(32) = \underline{736}$. The Σxy means to sum the products of each pair of values. $\Sigma xy = (1)(7) + (6)(8) + (9)(2) + (3)(5) + (4)(10) = 7 + 48 + 18 + 15 + 40 = \underline{128}$.

Occasionally functional values must be found for a certain set of x values. A function is simply a rule of correspondence (typically in the form of a formula) that describes exactly how each functional value is obtained from each of the various x values. For example, consider the function $F(x) = x/10$, for $x = 1, 2, 3, 4$. The definition of this function states that the functional value assigned to any one of the x's (1, 2, 3, and 4) is found by dividing that x by 10. That is, when $x = 1$, the value of $F(x)$ is 1/10 (usually written: $F(1) = 1/10$). Table I-1 shows the values of x and $F(x)$.

Table I-1

x	F(x)
1	1/10
2	2/10
3	3/10
4	4/10

Implied in this kind of functional definition is the fact that $F(x)$ does not exist for x values different from 1, 2, 3, and 4. [That is, $F(5)$ is not defined, $F(1.3)$ is not defined, and so on.]

Illustration Suppose that $Q = \Sigma [x \cdot F(x)]$. How would we obtain the value of Q, where $F(x)$ is defined as above?

Solution Q is equal to the sum of the products of each value of x with its related $F(x)$. Thus

$$Q = (1 \times 1/10) + (2 \times 2/10) + (3 \times 3/10) + (4 \times 4/10)$$

$$= \frac{1 + 4 + 9 + 16}{10} = \frac{30}{10} = \underline{3.0}$$

7 RANDOM NUMBERS Lesson I-3

EXERCISES I-2 (CONTINUED)

4. Given that $T = \Sigma x$, where $x = 1, 3, 2, 3, 5, 7$:
   a. find T;
   b. if $n = 6$, find $\bar{x}$ when $\bar{x} = \Sigma x/n$;
   c. find Σx^2;
   d. find $(\Sigma x)^2$;
   e. find the value of M if $M = n(\Sigma x^2) - (\Sigma x)^2$.

5. If $G(x) = x/6$ for $x = 1, 2, 3$, find:
   a. $G(1)$;
   b. $\Sigma[G(x)]$;
   c. $\Sigma[x \cdot G(x)]$;
   d. $\Sigma[x^2 \cdot G(x)]$.

6. Given $n = 5$ and the table of values shown here:

x	2	5	3	2	4
y	3	7	1	4	5

   Find each of the following expressions.
   a. Σx
   b. Σy
   c. Σxy
   d. $n(\Sigma xy) - \Sigma x \Sigma y$
   e. Σx^2
   f. Σy^2
   g. $\Sigma(x + y)$

LESSON I-3: RANDOM NUMBERS

The random-number table is a collection of random digits. The term "random" means that these digits are so arranged that each digit has an equal probability of occurrence. The digits presented in Table 1, page 566 of the textbook, are arranged in pairs and grouped into five rows and five columns. This format is used only for convenience. Other tables of random digits may be arranged differently.

Random numbers are used in two basic ways: (1) to select (identify) the source elements of a population that are to be used in a random sample, or (2) to simulate an experiment. When using the table of random digits, one may use them as single-digit numbers (0 to 9), as two-digit numbers (00 to 99), or as numbers of any desired size (000 to 999, etc.). This is accomplished by reading across the columns in the table to obtain the desired number of digits.

One of the uses of random numbers is to select a random sample. A simple random sample may be obtained by placing the identity of all elements in the population on tags or in capsules, putting them in a container, mixing thoroughly, and then drawing the sample from the container in lottery style. This method is often impractical due to the size of a population. For example, let's say that we are interested in drawing a sample from a population that contains 7564 people. These seven thousand plus people can be listed (any order) and numbered in succession (from 0001 to 7564). Once this list is numbered, you are ready to draw your sample (that is, to identify the sources for your sample data).

Sample data are identified by picking a starting point in your random-number table (or tables) and deciding on a pattern to follow in order to obtain enough identification numbers. The starting point is picked arbitrarily (perhaps by looking away and pointing to a position in the table). Turn to the table in your textbook and locate this sequence of digits: 26 25. (They are located in the fourth block of numbers to the right, six blocks down.) The pattern decided on is arbitrary, but once established it must be followed. This time we will go down the columns; when the bottom is reached we will start at the top of the next column to the right. We will continue in this manner until a large enough sample is identified. Notice that the first sources identified are 2625, 6196, and 5469. The fourth number is 7797; since there is no one on our list of 7564 people with that identification number, we will just ignore the 7797. 1302 is the next number, and we continue in this way until we obtain enough for our sample. In addition to occasionally finding numbers of no meaning (7565 to 9999, and 0000), we might on occasion find a number repeated. In that case, we would ignore the repetition as it would make no sense to use the same source twice in our sample.

EXERCISES I-3

1. Select the people to be used for a sample of size 12 in the above illustration.
2. A random sample is to be drawn from a list of 395 possibilities. Select the sources to be used if the list has been numbered from 001 to 395. Start at the lower right-hand corner of the table in the textbook and go up the column containing the last three digits. Identify ten.

The other main use of random numbers is to simulate a probability experiment. This simulation is accomplished by assigning numbers to each of the possible outcomes of a particular experiment. This assignment must be done in such a way as not to change the probabilities associated with each of the possible outcomes.

To illustrate this assignment, let's consider the experiment of tossing a coin several times and observing a head or tail on each toss. The probability of observing heads or tails is exactly 1/2. Suppose we let the occurrence of an even digit indicate the occurrence of a head, an odd digit a tail. That is, H = {0,2,4,6,8} and T = {1,3,5,7,9}. The probability of obtaining an even single-digit number is 1/2; the probability of obtaining an odd digit is also 1/2. (Other assignments could be used also.)

Once this assignment of numbers is made, one needs to decide on a procedure (path to be followed) for use of the table and pick a starting point. Then by following that procedure, each observed number indicates a particular occurrence, and the sample is quickly taken. Just remember that the probabilities must always be preserved. In our coin-tossing experiment, to simulate fifty tosses we could observe fifty single-digit numbers from the random-number table. These could be obtained as described in selecting a random sample, or a "block" of fifty digits could be used. Count the number of even digits and you will know how many heads have occurred from this simulated experiment.

EXERCISES I-3 (CONTINUED)

3. How many heads would occur in the coin-tossing simulation if you used the "50 block" in the upper left-hand corner of the random-number table?
4. Starting with the ninth digit down in the tenth column of single digits and proceeding down the column, what are the results of flipping a coin ten times?
5. Simulate the rolling of a pair of dice by using two columns of single digits where the first column represents one die and the second represents the other. The single digits 1, 2, 3, 4, 5, and 6 represent those values and 0, 7, 8, and 9 are meaningless. When one or both of the two digits are meaningless, the pair will be discarded. Start with 86 at the top of the tenth double column, go down the column, and simulate ten rolls of a pair of dice. Record your results as pairs of digits.

LESSON I-4: THE COORDINATE-AXIS SYSTEM AND THE EQUATION OF A STRAIGHT LINE

The rectangular coordinate-axis system is a graphic representation of points. Each point represents an ordered pair of values. (Ordered means that when values are paired, one value is always listed first, the other second.) The pair of values represents a horizontal location (the x-value, called the abscissa) and a vertical location (the y-value, the ordinate) in a fixed reference system. This reference system is a pair of perpendicular real number lines whose point of intersection is the zero of each line (Figure I-2).

Any point (x,y) is located by finding the point that satisfies both "positional" values. For example, the point $P(2,3)$ is exactly 2 units to the right of zero along the horizontal axis and 3 units above zero along the vertical axis. Figure I-3 shows two lines, A and B. Line A represents all the points that are 2 units to the right of zero along the x-(horizontal) axis. Line B represents all the points that are 3 units above the zero along the y-(vertical) axis. Point P is the one point that satisfies both conditions. Typically we think of locating point P by moving along the x-axis 2 units in the positive direction and then moving parallel to the y-axis 3 units in the positive direction (Figure I-4).

If either value is negative, we just move in a negative direction a distance equal to the number value.

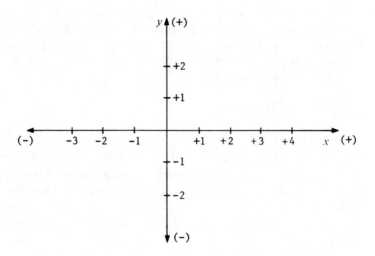

Figure I-2

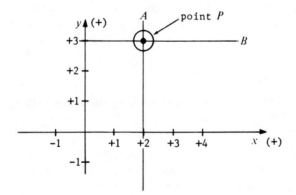

Figure I-3

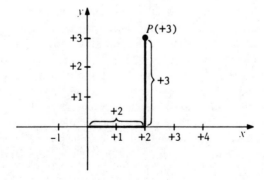

Figure I-4

COORDINATE-AXIS SYSTEM Lesson I-4

EXERCISES I-4

1. On a rectangular coordinate axis, drawn on graph paper, locate the following points.

$A(5,2)$ $B(-5,2)$ $C(-3,-2)$
$D(3,0)$ $E(0,-2)$ $F(-2,5)$

The equation of a line on a coordinate-axis system is a statement of fact about the coordinates of all points that lie on that line. This statement may be about one of the variables or about the relationship between the two variables. In Figure I-3 a vertical line was drawn at $x = +2$. A statement that could be made about this line is that every point on it has an x-value of 2. Thus, the equation of this line is $x = 2$. The horizontal line that was drawn on the same graph passed through all the points where the y-values were $+3$. Therefore the equation of this line is $y = +3$.

All vertical lines will have an equation of $x = a$, where a is the value of the abscissa of every point on that line. Likewise, all horizontal lines will have an equation of $y = b$, where b is the ordinate of every point on that line.

In statistics, straight lines that are neither vertical nor horizontal are of greater interest. Such a line will have an equation that expresses the relationship between the two variables x and y. For example, it might be that y is always one less than the double of x; this would be expressed equationally as $y = 2x - 1$. There are an unlimited number of ordered pairs that fit this relationship; to name a few: $(0,-1), (1,1), (2,3), (-2,-5), (1.5,2), (2.13,3.26)$ (see Figure I-5).

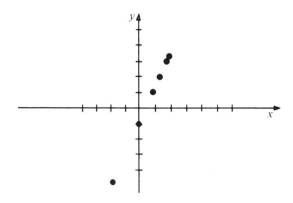

Figure I-5

Notice that these points fall on a straight line. Many more points could be named that also fall on this same straight line; in fact, all the pairs of values that satisfy $y = 2x - 1$ lie somewhere along it. The converse is also true: all the points that lie on this straight line have coordinates that make the equation $y = 2x - 1$ a true statement. When drawing the line that represents the equational relationship between the coordinates of points, one needs to find only two points of the straight line; however, it is often useful to locate three or four to ensure accuracy.

EXERCISES I-4 (CONTINUED)

2. a. Find the missing values in the accompanying chart of ordered pairs, where $x + y = 5$ is the relationship.

x	y
-2	
0	
	4
3.5	
	-1

b. On a coordinate axis, locate the same five points and then draw a straight line that passes through all of them.

3. Find five points that belong to the relationship expressed by $y = (3x/2) - 4$; then locate them on an axis system and draw the line representing $y = (3x/2) - 4$.

The form of the equation of a straight line that we are interested in is called the slope-intercept form. Typically in mathematics this slope-intercept form is expressed by $y = mx + b$, where m represents the concept of "slope" and b is the "y-intercept."

Let's look at the y-intercept first. If you will look back at Exercise 3, you will see that the y-intercept for $y = (3x/2) - 4$ [$y = (3x/2) + (-4)$] is -4. This value is simply the value of y at the point where the graph of the line intersects the y-axis, and all nonvertical lines will have this property. The value of the y-intercept may be found on the graph or from the equation. From the graph it is as simple as identifying it, but we will need to have the equation solved for y in order to identify it from the equation. For example, in Exercise 2 we had the equation $x + y = 5$; if we solve for y, we have $y = -x + 5$, and the y-intercept is 5 (the same as is found on the graph for Exercise 2).

The *slope* of a straight line is a measure of its inclination. This measure of inclination can be defined as "the amount of vertical change that takes place as the value of x increases by exactly one unit." This amount of change may be found anywhere on the line since this value is the same everywhere on a given straight line. If we inspect the graph drawn for Exercise 2 (Figure I-6), we will see that the slope is -1, meaning that each x increase of one unit results in a decrease of one unit in the y-value.

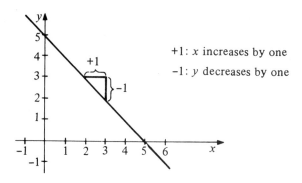

Figure I-6

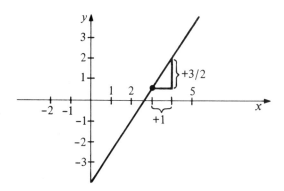

Figure I-7

An inspection of the graph drawn for Exercise 2 (Figure I-7) will reveal a slope of +3/2. This means that y increased by 3/2 for every increase of one unit in x.

The slope (m) may also be defined as the ratio of the change in y (Δy) to the change in x (Δx) as we move from one point to another along the straight line. Thus, $m = \Delta y / \Delta x$.

Illustration Find the slope of the straight line that passes through the points $(-1,1)$ and $(4,11)$.

Solution $$m = \frac{\Delta y}{\Delta x} = \frac{11 - 1}{4 - (-1)} = \frac{10}{5} = \underline{2}$$

Notes: 1. $\Delta y = y_2 - y_1$ and $\Delta x = x_2 - x_1$, where (x_1, y_1) and (x_2, y_2) are the two points that the line passes through.

2. As stated before, these properties are both algebraic and graphic. If you know about them from one source, then the other must agree. Thus, if you have the graph, you should be able to read these values

from it, or if you have the equation, you should be able to draw the graph of the line with these given properties.

Illustration Graph $y = 2x + 1$.

Solution $m = 2$ and $b = 1$. Locate the y-intercept ($y = +1$) on the y-axis. Then draw a line that has a slope of 2 (Figure I-8).

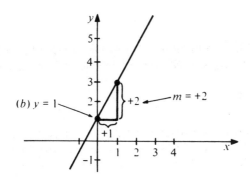

Figure I-8

Illustration Find the equation of the line that passes through $(-1,1)$ and $(2,7)$.

Solution (Algebraically) $m = \dfrac{\Delta y}{\Delta x} = \dfrac{7 - 1}{2 - (-1)} = \dfrac{6}{3} = 2$

$y = 2x + b$, and the line passes through $(2,7)$. Therefore $x = 2$ and $y = 7$ must satisfy (make the statement true) $y = 2x + b$. In order for that to happen, b must be equal to 3 $[7 = 2(2) + b]$. Therefore the equation of such a line is $y = 2x + 3$.

(Graphically) Draw a graph of a straight line that passes through $(-1,1)$ and $(2,7)$; then read m and b from it (Figure I-9).

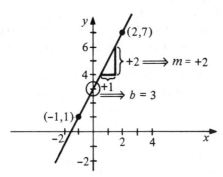

Figure I-9

Therefore $y = mx + b$ becomes $y = 2x + 3$.

TREE DIAGRAMS Lesson I-5

EXERCISES I-4 (CONTINUED)

4. Write the equation of a straight line whose slope is 10 and whose y-intercept is -3.

5. Draw the graph of each of the following equations (use graph paper).
   a. $y = x + 2$
   b. $y = -2x + 10$
   c. $y = (1/3)x - 2$

6. Find the equation of the straight line that passes through each of the following pairs of points (a) algebraically and (b) graphically (use graph paper).
   I. (3,1) and (9,5)
   II. (-2,3) and (6,-1)

The preceding discussion about the equation of a straight line is presented from a mathematical point of view. Mathematicians and statisticians often approach concepts differently. For instance, the statistician typically places the terms of a linear equation in exactly the opposite order from the mathematician's equation. To the statistician, for example, the equation of the straight line is $y = b + mx$, while to the mathematician it is $y = mx + b$. m and b represent exactly the same concepts in each case — the different order is a matter of emphasis. The mathematician's first interest in an equation is usually the highest powered term; thus he places it first in the sequence. The statistician tends to describe a relationship in as simple a form as possible; thus his first interest is usually the lower powered terms. The equation of the straight line in statistics is $y = b_0 + b_1 x$, where b_0 is the y-intercept and b_1 represents the slope.

LESSON I-5: TREE DIAGRAMS

The purpose of this lesson is to learn how to construct and read a tree diagram. A tree diagram is a drawing that schematically represents the various possible outcomes of an experiment. It is called a tree diagram because of the branching concept that it demonstrates.

Let's consider the experiment of tossing one coin one time. We will start the experiment by tossing the coin and will finish it by observing a result (heads or tails) (see Figure I-10).

```
Start          Observation
               Heads
               Tails
```

Figure I-10

16   BASIC MATHEMATICAL IDEAS   Introduction

This information is expressed by the "tree" shown in Figure I-11.

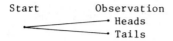

Figure I-11

In reading a diagram like this, the single point at the left is simply interpreted as, we are ready to start and do not yet know the outcome. The branches starting from this point must represent all of the different possibilities. With one coin there are only two possible outcomes, thus two branches.

EXERCISES I-5

1. Draw a tree diagram that shows the possible results from rolling a single die once.

2. Draw a tree diagram showing the possible methods of transportation that could be used to travel to a resort area. The possible choices are car, bus, train, and airplane.

Now let's consider the experiment of tossing a coin and a single die at the same time. What are the various possibilities? The coin can result in a head (H) or a tail (T) and the die could show a 1, 2, 3, 4, 5, or 6. Thus to show all the possible pairs of outcomes we must decide which to observe first. This is an arbitrary decision as the order of observation does not affect the possible pairs of results. To construct the tree to represent this experiment we list the above-mentioned possibilities in columns, as shown in Figure I-12.

```
    Start           Coin            Die

                                     1
                                     2
                     H               3
                                     4
                                     5
                                     6
                     T
```

Figure I-12

From "start" we draw two line segments that represent the possibility of H or T (Figure I-13). The top branch means that we might observe a head on the coin. Paired with it is the result of the die, which could be any one of the six numbers. We see this represented in Figure I-14.

17 TREE DIAGRAMS   Lesson I-5

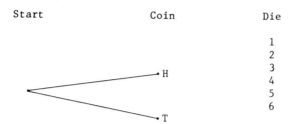

Figure I-13

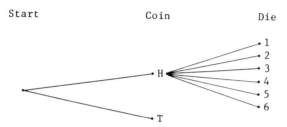

Figure I-14

However, the outcome could have been T, so T must have branches going to each of the numbers 1 through 6. To make the diagram easier to read we list these outcomes again and draw another set of branches (Figure I-15).

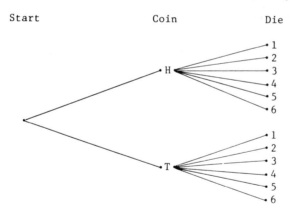

Figure I-15

Figure I-15 shows twelve different possible pairs of results — each one of the twelve branches on the tree represents one of these pairs. (A complete branch is a path from the start to an end.) The twelve branches in Figure I-15, from top to bottom, are H1, H2, H3, H4, H5, H6, T1, T2, T3, T4, T5, T6.

The ordering could have been reversed: the tree diagram in Figure I-16 shows the die result first and the coin result second.

18 BASIC MATHEMATICAL IDEAS   Introduction

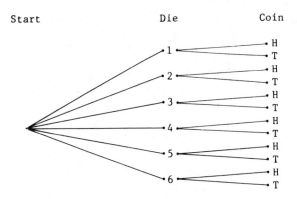

Figure I-16

Is there any difference in the listing of the twelve possible results? Only the order of observation and the vertical order of the possibilities as shown on the diagram have changed. They are still the same twelve pairs of possibilities. If the experiment contains more than two stages of possible events we may expand this tree as far as needed.

EXERCISES I-5 (CONTINUED)

3. a. Draw a tree diagram that represents the possible results from tossing two coins.
   b. Repeat part a considering one coin a nickel and the other a penny.
   c. Is the list of possibilities for part b any different than it was for part a?

4. Draw a tree diagram representing the tossing of three coins.

5. a. Draw a tree diagram representing the possible results that could be obtained when two dice are rolled.
   b. How many branch ends does your tree have?

On occasion the stages of an experiment will be ordered; when this is the case the tree diagram must show ordered sets of branches, as in Figure I-17. The experiment consists of rolling a die. Then the result of the die will dictate your next trial. If an odd number results, you will toss a coin. If a two or a six occurs, you stop. If any other number (a four) occurs, you roll the die again.

Notice that the tree diagram becomes a very convenient "road map" showing all the various possibilities that may occur in an experiment of this nature. Remember that an event is represented by a complete branch (a broken line from the start to an end) and the number of ends of branches is the same as the number of possibilities for the experiment. There are fourteen branches in the tree diagram in Figure I-17. Do you agree?

19   SETS AND SUBSETS   Lesson I-6

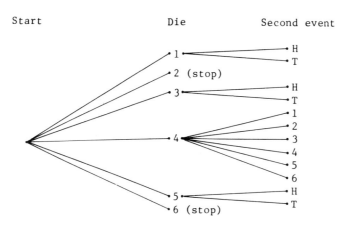

Figure I-17

EXERCISES I-5 (CONTINUED)

6. Students at our college are to be classified as to being male or female, graduates of public or private high schools, and as to the type of curriculum they are enrolled in: liberal arts or career. Draw a tree diagram which shows all of the various possible classifications.

7. There are two scenic routes (A and B) as well as one business route (C) by which you may travel from your home to a nearby city. You are planning to drive to that city by way of one route and come home by a different route.
   a. Draw a tree diagram representing all of your possible choices for going and returning.
   b. How many different trips could you plan?
   c. How many of these trips are scenic in both directions?

LESSON I-6: SETS AND SUBSETS

A set is a collection of items. The items that belong to the set are called elements or members of the set. A set must be carefully defined to enable us to determine its membership (that is, to be able to distinguish between members and nonmembers). The set that is of fundamental importance in statistics is the population. The population refers to the set of all source elements from which a particular piece of data, or information, is to be obtained. The population can also be thought of as the set of all values resulting from this set of source elements.

A subset consists of some part of the original set. In statistics, a subset is generally referred to as a sample of the population. One kind of subset that is often used is the complement, which occurs when we have a set (set A) and some subset (set B) of the given set in mind. The complement of the set B (identified by $\bar{B}$) is the set of all the elements that belong to set A and were not included in set B.

Note: The use of the bar ($\bar{A}$) for the complement set is unfortunate since the bar symbol will also be used for sample mean. There are three different symbols used in mathematics for the complement set, $\bar{A}$, $\tilde{A}$, and A'; all three symbols will be used in this textbook. In order to distinguish between complement set and sample mean, we will need to rely on the context and the usage of lowercase and capital letters. Capital letters are used for sets and lowercase letters for sample statistics.

Thus sets B and $\bar{B}$ ("B complement") partition the elements of set A into two nonoverlapping cells, yet the two cells contain all the elements of set A. A set whose subsets are being worked with is usually called the *universal set, U*. This universal set is the *population* in statistics and the *sample space* in probability: it is the set that contains all the elements that could possibly be included within a given situation.

To illustrate some of these ideas, let's consider the set of all digits to be the universal set, that is, $U = \{0,1,2,3,4,5,6,7,8,9\}$. Let set E be the set of even digits. Thus $E = \{0,2,4,6,8\}$, and therefore the complement of E, $\bar{E}$, is $\{1,3,5,7,9\}$. $\bar{E}$ could be described as the set of odd digits. Notice that there are no elements that belong to both E and $\bar{E}$ and that together E and $\bar{E}$ contain all of the elements of U. As another example, if you define set A to be $A = \{1,4,9\}$, then $\bar{A}$ is the set $\{0,2,3,5,6,7,8\}$.

There are two primary set operations that we must be familiar with: intersection and union. The *intersection* of two sets is very much like the intersection of two streets: it contains all elements that are common to the two (Figure I-18).

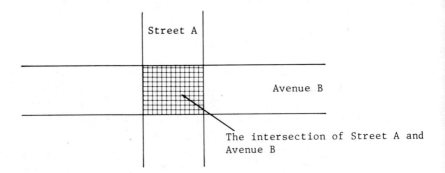

Figure I-18

Given sets $A = \{1,4,9\}$ and $E = \{0,2,4,6,8\}$, then the intersection of sets A and E (symbolized by $A \cap E$) is the set containing just the 4, $\{4\}$, since it is the only element that is common to both individual sets.

SETS AND SUBSETS Lesson I-6

The second operation is *union*, which might be thought of as the total of both sets combined. The union of Street A with Avenue B would be the surface to be repaired if both were to be repaired. With sets A and E, the union (written $A \cup E$) might be thought of as the set that would be formed if we were to dump both sets into the same container (Figure I-19). The result would be $A \cup E = \{0,1,2,4,6,8,9\}$. Notice that the receiving container ($A \cup E$) would have received two 4's. Repeated elements, however, are not listed twice, just as the highway department would not repair the intersection of Street A and Avenue B twice just because both streets were to be repaired.

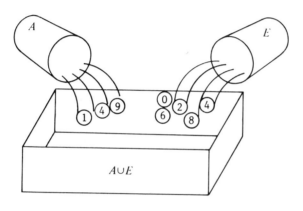

Figure I-19

Illustration   Find the complement of $A \cap E$.

Solution .   The complement of $A \cap E$ contains all those elements from the universal set that are not in $A \cap E$.

$$A \cap E = \{4\} \quad \text{thus} \quad \overline{(A \cap E)} = \{0,1,2,3,5,6,7,8,9\}$$

Illustration   Find the union of A complement with E.

$\overline{A}$ = $\{0,2,3,5,6,7,8\}$ $E = \{0,2,4,6,8\}$

$\overline{A} \cup E = \{0,2,3,4,5,6,7,8\}$

EXERCISES I-6

1. Given $U = \{0,1,2,3,4,5,6\}$, $B = \{0,1,2\}$, and $C = \{1,3,5\}$, find each of the following sets.
   a. $B \cap C$
   b. $\overline{B}$
   c. $B \cup C$
   d. $B \cup \overline{C}$
   e. $\overline{B} \cup C$
   f. $\overline{(B \cap C)}$
   g. $\overline{(B \cup C)}$
   h. $\overline{B} \cup \overline{C}$

2. A biology class includes the following students: Angela, Beth, Jim, Renee, and Norm, while a statistics class includes Bob, Steve, Beth, Russ, Jim, Pat, and Mack.

a. Find the intersection of these two classes and interpret the meaning of this intersection.
b. The union of these two classes is to participate in a field trip. Find the union of the two classes. How many people are in this union? Why is this number not 12?

3. Describe each of the following sets of students at our college.
    a. The intersection of the sets of students who are art majors and the set of students currently enrolled in statistics.
    b. The union of the set of black students with the set of foreign students.
    c. The intersection of the set of black students and the set of veterans.
    d. The union of the set of liberal arts majors with the set of students who have completed a statistics course.

LESSON I-7: VENN DIAGRAMS

The Venn diagram is a useful tool for representing sets. It is a pictorial representation that uses geometric configurations to represent "set containers." For example, a set might be represented by a circle — the circle acts like a "fence" and encloses all of the elements that belong to that particular set. The figure drawn to represent a set must be closed and the elements are either inside the boundary and belong to that set or they are outside and do not belong to that set. The universal set (sample space or population) is generally represented by a rectangular area, and its subsets are generally circles inside the rectangle. Complements, intersections, and unions of sets then become regions of various shapes as prescribed by the situation. The Venn diagram in Figure I-20 shows a universal set and a subset P.

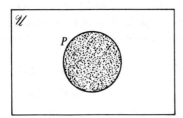

Figure I-20

Any element that is represented by a point inside the rectangle is an element of the universal set. Likewise, any element represented by a point inside the circle, P (the shaded area), is a member of set P. The unshaded area of the rectangle then represents $\bar{P}$.

The Venn diagrams in Figure I-21 show the regions representing $A \cap B$, $A \cup B$, $(\overline{A \cap B})$, and $(\overline{A \cup B})$. The shaded regions represent the identified sets.

VENN DIAGRAMS Lesson I-7

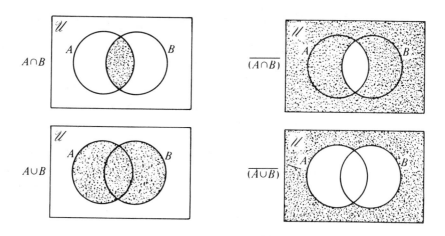

Figure I-21

When three subsets of the same population are being discussed, three circles can be used to represent all of the various possible situations.

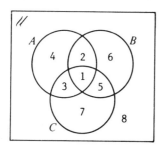

Figure I-22

In Figure I-22 the eight regions that are formed by intersecting the three sets have been numbered for convenience. Each of these regions represents the intersection of three sets (sets and/or complements of sets), as shown below.

Region Number	Set Representation
1	$A \cap B \cap C$
2	$A \cap B \cap \bar{C}$
3	$A \cap \bar{B} \cap C$
4	$A \cap \bar{B} \cap \bar{C}$
5	$\bar{A} \cap B \cap C$
6	$\bar{A} \cap B \cap \bar{C}$
7	$\bar{A} \cap \bar{B} \cap C$
8	$\bar{A} \cap \bar{B} \cap \bar{C}$

Region 1 ($A \cap B \cap C$) might be thought of as the set of elements that belong to A, B, and C. Region 4 represents the set of elements that belong to A, $\bar{B}$ (but not to set B), and $\bar{C}$ (but not to set C). Region 8 represents the set of elements that belong to $\bar{A}$, $\bar{B}$, and $\bar{C}$ (or that do not belong to A, B, or C). The others can be described in similar fashion.

Figure I-23 shows the union of sets B and C in the shaded areas of all three sets. Notice that $B \cup C$ is composed of regions 1, 2, 3, 5, 6, and 7. (You might note that three of these regions are inside A and three are outside.)

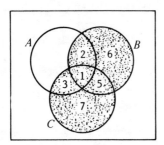

Figure I-23

EXERCISES I-7

1. Shade the regions that represent each of the following sets on a Venn diagram as shown in Figure I-24.
   a. A b. B c. $A \cap B$
   d. $A \cup B$ e. $\bar{A} \cup B$ f. $\bar{A} \cup \bar{B}$

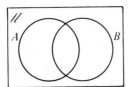

Figure I-24

2. On a diagram showing three sets, P, Q, and R, shade the regions that represent the following sets.
   a. P b. $P \cap Q$ c. $P \cup R$ d. $\bar{P}$
   e. $P \cap Q \cap R$ f. $P \cup \bar{Q}$ g. $P \cup Q \cup R$ h. $P \cup Q \cup \bar{R}$

LESSON I-8: THE USE OF FACTORIAL NOTATION

The factorial notation is a shorthand way to identify the product of a particular set of integers. 5! (five factorial) stands for the product of all positive integers

BINOMIAL COEFFICIENT   Lesson I-9

starting with the integer 5 and proceeding downward (in value) until the integer 1 is reached. That is, $5! = 5 \times 4 \times 3 \times 2 \times 1$, which is 120. Likewise, $n!$ symbolizes the product of the integer n multiplied by the next smaller integer $(n - 1)$ multiplied by the next smaller integer $(n - 2)$ and so on, until the last integer, the number 1, is reached.

Notes: 1. The number in front of the factorial symbol (!) will always be a positive integer or zero.
2. The last integer in the sequence is always the integer 1, with one exception: $0!$ (zero factorial). The value of zero factorial is defined to be 1, that is, $0! = 1$.
$1!$ (one factorial) is the product of a sequence that starts and ends with the integer 1, thus $1! = 1$.
$2!$ (two factorial) is the product of 2 and 1. That is, $2! = (2)(1) = 2$.
$3! = (3)(2)(1) = 6$
$n! = (n)(n-1)(n-2)(n-3)\ldots(2)(1)$
$(n-2)! = (n-2)(n-3)(n-4)\ldots(2)(1)$
$(4!)(6!) = (4 \cdot 3 \cdot 2 \cdot 1)(6 \cdot 5 \cdot 4 \cdot 3 \cdot 2 \cdot 1) = (24)(720) = 17280$
$4(6!) = (4)(6!) = 4(6 \cdot 5 \cdot 4 \cdot 3 \cdot 2 \cdot 1) = (4)(720) = 2880$

$$\frac{6!}{4!} = \frac{(6)(5)(4)(3)(2)(1)}{(4)(3)(2)(1)} = (6)(5) = 30$$

Table 2 in Appendix E on page 568 of the textbook gives the values of $n!$ for all integers of n from 0 to 20.

EXERCISES I-8

Evaluate each of the following factorials.
1. $4!$
2. $6!$
3. $8!$
4. $(6!)(8!)$
5. $\dfrac{8!}{6!}$
6. $\dfrac{8!}{4!\,4!}$
7. $\dfrac{8!}{6!\,2!}$
8. $2\,\dfrac{8!}{[5!]}$

LESSON I-9: THE BINOMIAL COEFFICIENT $\binom{n}{r}$

The binomial coefficient $\binom{n}{r}$ can be thought of as a number of different ways that a set of r items can be selected from a given set of n items. Let's look at a specific case first. $\binom{5}{2}$ is the symbol for the number of different ways that a set of two pencils can be selected from a set of five pencils. Suppose each pencil is a different color: red (R), blue (B), yellow (Y), green (G), and orange (O); thus picking yellow and green is distinctly different from picking red and yellow. However, to pick red and yellow is the same as to pick yellow and red. (The order of selection is of no consequence; only the set which results from the selection counts.)

The following list shows the set of all possible different selections: RB, RY, RG, RO, BY, BG, BO, YG, YO, GO. The list contains ten different possible choices, thus $\binom{5}{2}$ has the value of 10. Notice that for every set of two items that were selected, there was a set of three that were not selected; that is, if R and B were selected, then Y, G, and O were not selected; if RY were selected, then B, G, and O were not; and so on. This would suggest that the number of ways that exactly two items could be selected from among five items is exactly the same as the number of ways that five items could be separated into two sets, one set containing two items and the other three. But this number would have to be the same as the number of ways that exactly three could be selected from the set of five: RB,YGO; RY,BGO; RG,BYO; RO,BYG; BY,RGO; BG,RYO; BO,RGY; YG, RBO; YO, RBG; GO, RBY.

Does it matter whether you (1) split the sets into two parts, (2) select two items (and not the other three), or (3) select three items (and not the other two)? No, they're all the same value.

The symbol $\binom{n}{r}$ is evaluated by the formula

$$\binom{n}{r} = \frac{n!}{r!\,(n-r)!}$$

which in a sense shows three numbers: the total number of elements (n), the number being selected (r), and the number left over ($n - r$). The factorials are due to the number of possible selections left for each next step. Using the formula to evaluate $\binom{5}{2}$, we find

$$\binom{5}{2} = \frac{5!}{(2!)\,(5-2)!} = \frac{5 \cdot 4 \cdot 3 \cdot 2 \cdot 1}{(2 \cdot 1)\,(3 \cdot 2 \cdot 1)} = 10$$

EXERCISES I-9

1. Evaluate each of the following terms.
   a. $\binom{5}{4}$ b. $\binom{6}{3}$ c. $\binom{10}{4}$ d. $\binom{17}{15}$ e. $\binom{8}{5}$

2. Justify that $\binom{n}{k} = \binom{n}{n-k}$, where $k < n$ and both are integers.

BINOMIAL COEFFICIENT   Lesson I-9

3. Explain why it is necessary for both n and r to be integers if the symbol $\binom{n}{r}$ is to be meaningful.

Table 3 in Appendix E, page 569 in the textbook, shows the values for $\binom{n}{k}$ for all possible values of n and k up to 20. To avoid repetition, the values of k are read directly up to $k = 10$. For values of $k > 10$, you must use the statement in Exercise 2: $\binom{n}{k} = \binom{n}{n-k}$. Thus to find $\binom{15}{13}$ we would look up $\binom{15}{2}$.

EXERCISES I-9 (CONTINUED)

4. Check your answers for Exercise 1 by looking up the values in Table 3.

5. Find the value of $\binom{21}{5}$, using the values in Table 3. [Hint: Use the idea of Pascal's triangle and add two numbers together to find it.]

STATISTICS

1

LESSON 1-1: BASIC TERMS

You absolutely must understand certain basic terms in order to follow the discussions in the textbook. Eight of these terms will be explained in this lesson; however, I would suggest that you review their definitions on pages 6, 7, and 8 in the textbook before continuing here.

Now that you have refreshed your memory, let's take another look at each term individually.

The *population* is like the universal set in that it contains all the items or all the values that could be considered within the framework of a given question. A few examples of populations made up of items are these: (1) all children who live in a certain state, (2) all household pets belonging to the residents of a certain city, (3) all light bulbs manufactured in the United States today, and (4) all apples in a certain orchard. A population can also be a collection of values; for example: (1) the "weights" of the children who live in a certain state, (2) the kind of household pet (dog, cat, fish, and so on) belonging to the residents of a certain city, (3) the "length of life" of each light bulb manufactured in the United States today, and (4) the "circumference" of each apple in a certain orchard. The term "population" is properly applied to either the set of items (sources) or the set of all "values" assigned to these sources.

In describing the set of populations consisting of values, the response variable was also identified. The *response variable* is the characteristic about each item that is of interest: weight (pounds), kind (dog, cat, and so on), length of life (hours of burning time), and circumference (inches). As you can see, the variables may be quantitative in nature or they may be qualitative. The classification of variables is the topic of the next lesson (Lesson 1-2).

Because most populations are too large to study in their entirety, a sample is obtained. A *sample* is a subset of the population, a collection of objects or values that belong to it. The sample is smaller, by design, than the population and thus more manageable with respect to the statistical operations to be performed on it.

Once the sample has been obtained, we can identify its data. A *piece of data* (singular) is any one value of the response variable. For example, one apple is selected from the orchard and measured; its circumference, 8.57 inches, is a single piece of data. Again, one resident in a city says that his only household pet is a gerbil, thus "gerbil" would be a piece of data for that sample. When the entire set of values for the sample has been collected, that set becomes the *data* (plural). (Notice that the terms "sample" and "data" could mean exactly the same thing if the sample is thought of as the set of values.)

An *experiment* is any kind of a planned activity whose results yield a value of the response variable. It could be as simple as looking at a person and identifying the color of his hair, or it could be as complex as setting up and running a test to measure the life of light bulbs. The important point is that the activity is *planned*. (All data should be obtained after the planning has been completed.)

The principal objective of taking a sample is to help us describe the population from which the sample was taken. Suppose that we were interested in describing the apples in our orchard. We could convey the size of these apples (all the apples in the orchard) by stating their average size. Any value, or measure, that describes some characteristic of the population (such as average size of apples) is called a *parameter*. If we wanted to know the value of this particular parameter, we would most likely select a sample of the apples in the orchard and obtain an "average size" for those apples belonging to the sample. This value exactly describes this characteristic (average size) of all apples belonging to the sample. Any such value that has been observed (or obtained) from a sample is called a sample *statistic*. The sample statistic is then used as a basis for making an estimation of the population parameter.

The following exercises will help you think about and sort out the meaning of these eight terms.

EXERCISES 1-1

1. Excerpt from a news article:

    Poll Finds Many Fear Job Loss

    The fear of unemployment is very threatening to some segments of the public, as the poll found that almost one of five American workers worried about losing their job.

    Carefully describe the following items as they apply to the excerpt.
    a. the population　　b. the variable　　c. the sample
    d. the parameter　　e. the statistic

2. Excerpt from a news article:

    Pay Checks

    Residents of Alaska and the District of Columbia lead the country in average annual salaries, while South Dakota comes in last.... The Labor Department statistics were based on the average annual pay for workers covered by state and federal unemployment insurance programs.

Carefully describe the following items as they apply to the excerpt.
a. the population b. the variable c. the parameter

LESSON 1-2: TYPES OF VARIABLES

The response variable, or piece of data, in a statistics problem may be classified as one of two kinds: qualitative or quantitative. (See diagram on page 9 of the textbook.) The qualitative piece of data is typically a descriptive kind of value that might be used to sort or categorize the elements of the population. Usually the qualitative response variable is a "word" rather than a number: a manufacturer's name, a type, a color, a preference, and so on. We will call data of this qualitative nature *attribute data*. Identification numbers belong to this category, too; for example, student identification numbers, license numbers, year of birth, and year of manufacture are numerical response variables which are "attributes."

If a variable is not qualitative, then it is quantitative. The group of quantitative variables is usually divided into two subgroups: discrete and continuous. This separation is essential. As a student of elementary statistics, you may not come to realize the full significance of that statement, but you will recognize some of these distinctions.

A *continuous variable* is any variable that "could" take on all the real number values. This means that the value of the variable could be measured with as many decimal places of accuracy as needed to achieve the exact value. This kind of measurement becomes virtually impossible from a practical standpoint. For example, your weight is a continuous variable. Typically we talk about our weight to the nearest whole pound (perhaps nearest five pounds), but few people weigh "exactly" a whole number of pounds. You might say that your weight is 148 pounds, but most likely you don't weigh exactly 148.000 pounds — scales used for weighing people are not that accurate. 148 pounds usually means a weight that is somewhere between 147.5 pounds and 148.5 pounds.

Any variable that by its very nature cannot take on all possible real number values is called a *discrete variable*. This means that there are unattainable numerical values between each adjacent pair of attainable values of the variable. The number of cavities that a person has ever had might be such a variable. When counting cavities it would be impossible to have any fractional values result. There are a few discrete variables that take on fractional or decimal values; however, the number of decimal values is limited (for example, just halves, or just tenths). Scores at a diving or figure-skating competition are examples of discrete variables since they can take on decimal values only to the tenths position but not the values in between the tenths.

TYPES OF VARIABLES   Lesson 1-2

You will nearly always be able to distinguish between continuous and discrete variables simply by asking yourself if the variable is a "measurement" of something or a "count" of items. A measurement will always be a continuous variable, a count will always be a discrete variable — when you count items, they either exist or they do not (a whole or nothing).

Of course you must be aware of misleading phrases: "number of miles" sounds like a count, but in fact it is a measurement (thus continuous). There are several variables that can be converted from one classification to another either by the way the information is used or by the type of information that is used. The manufacturer's "year" identification of an automobile (such as a "1979 model") is actually an attribute; however, this information is usually interpreted as the variable "age." Age of an automobile in this context would be a discrete variable since it would always be a whole number. But the "age of a car" could also be number of miles driven or age in terms of years and fractional parts of years that you have actually owned it. Both of these latter illustrations would be continuous.

Although attribute data are very useful, they yield us the least amount of information; thus there are fewer applications for them than for quantitative data. The descriptive statistics presented in Chapter 2 apply to quantitative data, with little distinction made between discrete and continuous variables. The significance of this distribution will become evident later, in the discussion on probability distribution and the theory of application.

EXERCISES 1-2

1. Classify each of the following variables as to: (A) *attribute* (qualitative information), (B) *discrete* (quantitative information), or (C) *continuous* (quantitative information).
    a. The first semester's academic average for a student enrolled this year at a high school in our county.
    b. The number of students per homeroom in high school A with an honor roll average.
    c. The number of students in our local elementary school that are on the safety patrol.
    d. The number of minutes that it takes a student in statistics to complete Lesson 1-2 in this study guide.
    e. The number of cracked eggs per dozen found on the shelf at a local grocery.
    f. The number of station wagons sold to new-car buyers in Monroe County during 1978.
    g. The number of "shocks" that each laboratory mouse receives before it completes the desired task.

SELF-CORRECTING EXERCISES FOR CHAPTER 1

1. Excerpt from a news article:

   Bar Survey Shows Lawyers Average $47,204 in 1983

   The information obtained last fall from 206 questionnaires ... showed that the average attorney ... earned $47,204. ...

   In reference to the excerpt, carefully describe or identify each of the following terms.
   a. population b. sample c. variable
   d. parameter e. statistic f. experiment

2.

   In reference to Benny's worry, carefully describe the following terms.
   a. population b. variable c. parameter

3. Excerpt from a news article:

Families Are Smaller

Families are only half as large as they were when our country was young. . . . In 1800 the average woman had seven children, five of whom survived; today she has two, and both survive.

Suppose you were asked to collect a sample of 100 to verify that today's average woman has two children. Carefully describe the following terms in reference to that data.

a. population   b. variable   c. sample
d. parameter   e. statistic

DESCRIPTIVE ANALYSIS AND PRESENTATION OF SINGLE-VARIABLE DATA

2

LESSON 2-1: THE FIVE "AVERAGES"

The "average" is a measure of central tendency that we are all familiar with. But is it really? When you read a statement like, "The average wage earned is $156," do you receive the message that was intended? When most of us hear the word "average," we tend to think of the number found by $\Sigma x/n$, or the mean, as it is correctly named. The value stated may or may not be the mean. There are five values that can be used as the "average." The value cited should be the one that presents the best possible picture of what the average situation really is. The best way to identify the kind of average used is to call it by its proper name. Each of these averages (mean, median, mode, midrange, and midquartile) is defined and described in Chapter 2 of the textbook. In this lesson we should like to present some illustrations of when and why each of the various "averages" is used.

The *mean* ($\bar{x}$) is the average value that is obtained by adding all the pieces of data (x) together and dividing by the number of pieces of data (n): $\bar{x} = \Sigma x/n$. Each piece of data plays an equal role in determining the mean. There are hundreds of everyday applications for the mean: your exam average, the average height of the members of the varsity basketball team, the average age of automobiles owned by students enrolled in statistics are three examples of means.

The *median* ($\tilde{x}$) is the value of the one piece of data (or the value halfway between the two pieces of data) in the middle of a set of data when the values are ranked in order according to size (smallest to largest). To find the median you need to know all the individual values, but the middle one alone determines the median. The median is the "average" most typically used in the reporting of salaries.

The *mode* is the one numerical value among the data that occurs most frequently. In a situation where one value appears most often, the mode will be the average to apply. However, the mode will not be applicable when there are two or more values that occur most frequently. The mode would be the average that you would want to use if you were a buyer for a clothing store: the most typical size is the one that you will sell the most of.

The *midrange* is the best average to use when the highest and lowest values are readily available: it is the value exactly midway between them. The most common use of the midrange is in weather reporting — the average daily temperature is actually the midrange of the daily high and low temperatures.

If you need to report the measure of central tendency of a particular situation, just remember to choose the average which best represents the message that you are trying to convey.

EXERCISES 2-1

In each of the following exercises, select and calculate the value of the measure of central tendency that seems to be the most representative. Explain.

1. A survey was taken as cars drove by a check point and the number of persons riding in each car was observed. The data were as follows: 12 cars had 1 person, 6 cars had 2 people, 4 cars had 3 people, 2 cars had 4 people, and 1 car had 6 people. How many passengers were in the "average" car?

2. The town assessor wishes to describe the value of houses in a certain area to the town board, which is studying the town's tax source. The homes are valued (in $1,000) as follows: 26, 29, 35, 27, 28, 29, 37, 31, 31, 27, 32, 31. What "average" is most appropriate for the assessor to use?

3. Explain why the median would sometimes serve as an "average."

LESSON 2-2: THE STANDARD DEVIATION OF A SAMPLE

The *standard deviation* of a set of data is a measure of the dispersion (spread) that the individual pieces of data display about their mean value. If the data are tightly grouped, the measure of dispersion will be quite small, and inversely, if the data are widely spread, then the measure of dispersion will be quite large.

Before we get to the "standard" deviation about the mean, we first need to be sure we understand what the term "deviation" means. To deviate means to stray from the standard. In statistics this standard is the mean value. And the measure of the "stray" is the distance that a given piece of data lies from the mean. Within a set of data (a sample), the mean value is identified by $\bar{x}$ ("x bar"), and each piece of data will deviate from the mean by a value found by subtracting the value of $\bar{x}$ from the value of each piece of data. Thus, any one deviation is represented by $(x - \bar{x})$.

Now what we would like to obtain is some kind of "average" value for all of these deviations. There are n deviations, one for each piece of data in the sample. Some deviations are negative, some are positive, and some might be zero — see page 58 of the text. As discussed in the textbook on pages 58-60, there is a problem when trying to find this "average value." The problem is that the sum of these deviations will always be exactly zero $[\Sigma(x - \bar{x}) = 0]$. Thus, to sum them up and divide by n is of no value, since we will always obtain zero. This is due to the "canceling" effect of the positive and negative deviations when added together. There are two ways to avoid this canceling effect: (1) use absolute value (see the discussion in the textbook on page 58), and (2) use the square of each deviation. Either of these methods will guarantee us a positive total.

The standard deviation uses the second method for obtaining a "total" of individual deviations. This total is then divided by an appropriate number in order to cause the "averaging effect." However, the resulting number has been affected by the squaring that took place earlier, so it seems quite proper to take the square root of the previously found "average value."

It would seem that we should divide by n whenever the "averaging" is done. This is exactly what we would do if the set of data were the *entire population*; however, in practice we need to divide by $n - 1$. Think of this as a fudge factor, if you wish, but $n - 1$ is the correct value to use when the set of data is a sample. (The data will almost always be a sample.) It is not possible at this level to justify the use of $n - 1$ in place of n.

The *standard deviation* of a sample of data, identified by s, is therefore defined by the formula:

$$s = \sqrt{\frac{\Sigma(x - \bar{x})^2}{n - 1}}$$ [formula (2-8), p. 61 in textbook]

The number under the radical sign in formula (2-8) is called the *variance* and is identified by s^2.

$$s^2 = \frac{\Sigma(x - \bar{x})^2}{n - 1}$$ [formula (2-7), p. 59 in textbook]

The term "variance" can be thought of as an intermediate value obtained along the way as the standard deviation is being calculated. Owing to the complex nature of the formula for the standard deviation, the formula for variance is often used as an intermediate step only to help simplify the standard-deviation formula.

The following illustration shows the use of formula (2-8) to obtain the standard deviation of a given sample.

Illustration Find the standard deviation of the sample 2, 3, 6, 4, 1, 2.

Solution In order to use formula (2-8), we will first need to find the sample mean $\bar{x}$.

$$\bar{x} = \frac{\Sigma x}{n} = \frac{2 + 3 + 6 + 4 + 1 + 2}{6} = \frac{18}{6} = \underline{3.0}$$

The data and formula are best organized in tabular form (Table 2-1).

STANDARD DEVIATION OF SAMPLE   Lesson 2-2

Table 2-1

x	$(x - \bar{x})$	$(x - \bar{x})^2$
2	-1	1
3	0	0
6	3	9
4	1	1
1	-2	4
2	-1	1
	0 (ck)	16

Note: $\Sigma(x - \bar{x}) = 0$, as stated earlier.

$$s = \sqrt{\frac{16}{5}} = \sqrt{3.20}$$

$$s = 1.788 = \underline{1.8}$$

EXERCISES 2-2

1. Find the standard deviation for each of the following samples.
   a. 7,2,3,4,3,5
   b. 1,2,5,8,4,3,7,2
   c. 10,8,11, 19, 7, 14, 12, 16, 12, 11
   d. 2, 3, 4, 3, 5

As part d of Exercise 1 might suggest, the formula for the standard deviation (2-8) could become very unwieldy if the mean does not happen to be a "nice" number to work with. (To round off the mean value would give an approximate value for the standard deviation and thus should not be used unless several extra decimal places are used.) Therefore, the "definition formula" (2-8) has been rewritten into what is often referred to as the "shortcut formula." (It is not always the shortest way to do the calculation, but it does bypass the need for the mean.) This so-called shortcut formula is:

$$s = \sqrt{\frac{\Sigma x^2 - \frac{(\Sigma x)^2}{n}}{(n - 1)}}$$ [formula (2-10), p. 62 in textbook]

In using this formula, one squares the values of the data, not the deviations. This is often quicker because your data will be numbers whose squares you know. (The squares of the deviations are usually unknown because of the decimals.) The solution to the following illustration uses both techniques.

Illustration   Find the standard deviation of the sample of data 1, 2, 2, 3, 5, 6.

Solution   [using formula (2-8)]

$$\bar{x} = \frac{\Sigma x}{n} = \frac{19}{6} = 3.166\overline{6} = 3.17$$

Table 2-2

x	$(x - \bar{x})$	$(x - \bar{x})^2$
1	-2.17	4.7089
2	-1.17	1.3689
2	-1.17	1.3689
3	-0.17	0.0289
5	1.83	3.3489
6	2.83	8.0089
	-0.02*	18.8334

*Note: This total is not quite zero due to the rounding off that took place.

$$s = \sqrt{\frac{18.8334}{5}} = \sqrt{3.76668}$$

$$s = 1.9408 = \underline{1.9}$$

Solution [using formula 2-10)]

Table 2-3

x	x^2
1	1
2	4
2	4
3	9
5	25
6	36
19	79

$$s = \sqrt{\frac{79 - \frac{(19)^2}{6}}{5}} = \sqrt{\frac{79 - 60.166}{5}} = \sqrt{\frac{18.833}{5}}$$

$$s = \sqrt{3.7666} = 1.9407 = \underline{1.9}$$

Notice that the answers for both solutions are identical. They will always be identical as long as the arithmetic is accurate and the round-off error is controlled. Look at the two solutions again. Which one would you prefer to use in a similar situation?

How do you decide which formula is the most appropriate to use? In a typical problem situation you will be asked to find the mean, so find it first. If the mean is a whole number then the definition formula, (2-8), will usually be the easiest

MEDIAN, QUARTILES, PERCENTILES   Lesson 2-3

to use. If the mean is a fraction or decimal, then the shortcut formula, (2-10), will usually be handiest.

EXERCISES 2-2 (CONTINUED)

2. Find the standard deviation for each of the following sets using the shortcut formula.
   a. 7, 5, 8, 9, 7
   b. 9, 8, 12, 11, 14, 9, 10, 12
   c. 4, 3, 2, 9, 6, 7, 5, 6, 4, 5

LESSON 2-3: LOCATING THE MEDIAN, THE QUARTILES, AND THE PERCENTILES

When locating one of the measures of position (the median, one of the quartiles, or one of the percentiles), the procedure is always the same: (1) rank the data; (2) determine the position number i (or depth d ()) for the statistic being sought; (3) determine the value of the statistic.

The position number i is like a street number for a house. It simply locates the house on the street. When a set of data is ranked (lowest to highest is our usual procedure), the smallest-valued data (L) has the location $i = 1$; the next-to-smallest-valued data has the location $i = 2$; and so on up to the largest-valued data, which has the location $i = n$. The position number i will either be an integer or it will be a number containing a fraction of 1/2. When i is an integer, the value of the statistic is the value of the data that occupies the ith position in our ranked data. When i contains the fraction 1/2, the value of the statistic is midway between the values of two adjacent data. For example, if $i = 7.5$, then the value of the statistic being sought is midway between the value of the 7th and the 8th data. The value midway between two numbers is found by adding the two values and dividing by two. This procedure will be demonstrated in the following illustrations.

Illustration   Consider the following sample data:

x	5	8	9	10	11	12	13	14	15	16	17	19	20	22	24	25	27	29	30	33
i	1	2	3	4	5	6	7	8	9	10	11	12	13	14	15	16	17	18	19	20

Find the value of x where $i = 5$.

Solution $x = \underline{11}$

Find the value of x where $i = 14.5$.

Solution

$$x = \frac{22 + 24}{2} = \underline{23}$$

Now let's look at the procedure for finding i for the measures of position.

Median. To find i, we use $i = (n + 1)/2$, where n is the number of pieces of data in the sample.

Quartiles. The quartiles are determined by the same procedure we use for percentiles. Remember that $Q_1 = P_{25}$ and $Q_3 = P_{75}$.

Percentiles. The procedure for finding i for the percentiles is a two-step procedure. First you must find the value of $nk/100$, where n is the sample size and k is the percentile being sought. For example, let's determine the value of $nk/100$ for P_{30} when $n = 50$.

$$n = 50 \quad \text{and} \quad k = 30$$

Therefore,

$$\frac{nk}{100} = \frac{(50)(30)}{100} = \underline{15.0}$$

Also let's determine the value of $nk/100$ for P_{24} when $n = 72$.

$$n = 72 \quad \text{and} \quad k = 24$$

Therefore,

$$\frac{nk}{100} = \frac{(72)(24)}{100} = \underline{17.28}$$

As you can see, the value of $nk/100$ may be either an integer (does not contain a fraction) or it may not be an integer (does contain a fraction).

Now you are ready to find i. If $nk/100$ is an integer, then i is equal to $(nk/100) + (1/2)$. For example, $nk/100 = 15.0$ above; therefore, $i = 15.0 + (1/2)$ or 15.5. On the other hand, if $nk/100$ is not an integer, then i is equal to the next larger integer. For example, $nk/100 = 17.28$ in one of the examples above. The corresponding i is then 18. (Note that no matter what the fraction value is, the value of i is the next larger integer, not the nearest integer.)

Illustration Find the value of P_{90} for the sample given in the previous illustration.

Solution $n = 20$ and $k = 90$; therefore,

$$\frac{nk}{100} = \frac{(20)(90)}{100} = \underline{18.0}$$

Thus $i = 18.50$ and

MEDIAN, QUARTILES, PERCENTILES Lesson 2-3

$$P_{90} = \frac{29 + 30}{2} = \underline{29.5}$$

Illustration Find the value of P_{62} for the sample in the first illustration of this section.

Solution $n = 20$ and $k = 62$; therefore,

$$\frac{nk}{100} = \frac{(20)(62)}{100} = 12.40$$

Thus $i = 13$ and

$$P_{62} = \underline{20}$$

Illustration Find the values of Q_1 and Q_3 for the sample of 20 pieces of data given in the first illustration of this section.

Solution For Q_1 we have $n = 20$ and $k = 25$; therefore, $nk/100 = 5.0$. Thus $i = 5.5$ and

$$Q_1 = \frac{11 + 12}{2} = \underline{11.5}$$

For Q_3 we have $n = 20$ and $k = 75$; therefore, $nk/100 = 15.0$. Thus $i = 15.5$ and

$$Q_3 = \frac{24 + 25}{2} = \underline{24.5}$$

An optional shortcut. The value of any quartile or percentile can also be found by counting down from the highest-valued data. The procedure is the same as described above except that k is replaced with $(100 - k)$. For example, the value of Q_3 for our illustration above is in the position $i = 5.5$ counting from H, the largest-valued data. The depth, d (), notation is the position relative to the extreme value, whether it be the lowest-valued or highest-valued data.

x	$\cdots$ 22 24 25 27 29 30 33
i	$\cdots$ 14 15 16 17 18 19 20
Counting down	(6) (5) (4) (3) (2) (1)

$$Q_3 = \frac{24 + 25}{2} = \underline{24.5}$$

This is the same value that we found previously.

EXERCISES 2-3

1. Find the value of i that would be used for locating each of the following in a sample of 40 data.

a. P_{10} b. Q_1 c. P_{38}

2. Make a list of the numbers from 1 to 50 along a line as though these were the position numbers of 50 pieces of data ranked according to size; the smallest is number 1 and the largest is number 50.
   a. Determine the position of the median.
   b. Determine the position of the third quartile.
   c. Renumber these 50 values from largest to smallest (50 becomes 1, 49 becomes 2, ..., 1 becomes 50). Find the median and the third quartile working from the highest-valued data. Compare these results to those in parts a and b.

3. Show that the value of $i = (n + 1)/2$ used for finding the median and the value of i found for the 50th percentile are exactly the same.

4. Repeat Exercise 2 with 75 pieces of data.

LESSON 2-4: FREQUENCY DISTRIBUTIONS

A *frequency distribution* is a listing of frequencies, and frequencies are numbers that indicate how many times a particular value has occurred within a sample. Illustration 2-1 on page 28 in the textbook shows a listing of 20 test scores. The stem-and-leaf-display in Figure 2-3 on page 29 shows the same set of 20 test scores. These scores could be listed in a frequency distribution since many of them occurred more than once. Table 2-4 shows the 20 scores as a frequency distribution. The f stands for frequency and the x is the value of the variable (rating score). The 3 opposite the 74 indicates that 3 scores were 74. The 20 at the bottom represents the sum of the frequencies (Σf) and is by necessity equal to the number of pieces of data (n) in the sample.

Table 2-4

x	f
52	1
58	1
62	1
66	1
68	1
72	1
74	3
76	2
78	2
82	2
84	1
86	1
88	1
92	1
96	1
	20

FREQUENCY DISTRIBUTIONS   Lesson 2-4

Illustration The advantage to using a frequency distribution is that it avoids a lot of unnecessary work. The following illustration demonstrates this.

Put five coins in a cup, shake, dump out, count the number of coins that landed heads up, and record; repeat 20 times.

Solution The resulting: 2 3 2 1 2 0 2 4 1 1
3 3 1 3 3 1 4 3 2 4

It's wasted effort to list these twenty values of x separately in order to work them. Listed in order from smallest to largest they look like this:

0 1 1 1 1 2 2 2 2 2 3 3 3 3 3 3 4 4 4

But these same twenty values can be shown a lot more compactly as a frequency distribution:

x	0	1	2	3	4
f	1	5	5	6	3

A more complete name for the frequency distribution discussed so far would be "ungrouped frequency distribution." It is ungrouped because each value of x listed is a value of x that occurred in the data, and every value that occurred in the data must be listed. (The frequency shows how many times it occurred.) The "grouped frequency distribution" is a listing of frequencies such as those listed beside class marks. The data in Table 2-4, which is an ungrouped frequency distribution, could be grouped and presented as a grouped frequency distribution as shown in Table 2-5.

Table 2-5

Class	Class mark	f
50–59	54.5	2
60–69	64.5	3
70–79	74.5	8
80–89	84.5	5
90–99	94.5	2
		20

The data (individual values) have been classified, and each class is assigned a value which stands for all values that fall into that class. This value is the "class mark" and it is used in much the same way as the value of x was used in the ungrouped frequency distribution — for convenience with large sets of data. But when the grouped distribution is formed, a certain amount of information is lost. Each individual value is treated as though it were the value of the class mark of the class to which it is assigned. This loss of identity results in easier computations and has only a slight rounding-off effect on the results.

DESCRIPTIVE ANALYSIS: SINGLE-VARIABLE DATA   Chapter 2

The main reason for the use of the frequency distribution (grouped or ungrouped) is to make the calculations more manageable when dealing with large sets of data. The application of the frequency distribution to calculations is the topic of the next lesson (Lesson 2-5).

EXERCISES 2-4

1. Display the following set of data as a frequency distribution.

20	25	30	30	20	25	35	20	40	15
35	5	25	15	20	20	20	25	25	30
25	20	20	20	25	20	25	5	20	20
15	25	20	15	30	10	45	25	30	15

2. Display the following set of data as an ungrouped frequency distribution.

1	6	3	4	2
3	4	5	2	4
3	9	4	3	3
3	5	2	3	7
4	3	5	1	3

3. Display the following set of data as a grouped frequency distribution. Use class limits of 60-62, 63-65, 66-68, ...

66	70	71	65	67	75	65	62	68	70
67	71	73	68	63	69	62	72	70	65
61	71	70	69	73	67	69	69	71	65
68	68	62	71	69	63	70	69	69	64
66	73	64	65	71	61	60	67	71	63

LESSON 2-5: CALCULATIONS USING A FREQUENCY DISTRIBUTION

In Lesson 2-4 we saw how a set of twenty values could be expressed more concisely by using a frequency distribution than by just listing them. One reason for introducing the frequency distribution was to aid us in the calculations. It seems a great waste of effort and time to list all twenty values of x and square each one as we did in Lesson 2-2 to calculate the mean and the standard deviation. To contrast and compare the two methods, let me calculate the standard deviation for the sample obtained in Lesson 2-4 for the number of heads observed when five coins are tossed. The values of x have been arranged in order according to size (Table 2-6).

FREQUENCY DISTRIBUTION CALCULATIONS Lesson 2-5

Table 2-6

x	x^2
0	0
1	1
1	1
1	1
1	1
1	1
2	4
2	4
2	4
2	4
2	4
3	9
3	9
3	9
3	9
3	9
3	9
4	16
4	16
4	16
Total 45	127

Shortcut Formula

$$s = \sqrt{\frac{\Sigma x^2 - \frac{(\Sigma x)^2}{n}}{(n-1)}}$$

$$s = \sqrt{\frac{127 - \frac{(45)^2}{20}}{(19)}}$$

$$s = \sqrt{1.355} = 1.164 = \underline{1.2}$$

As you can see in Table 2-6, when we sum the x column we add in one zero (0), five 1's (5), five 2's (10), six 3's (18), and three 4's (12). Wouldn't it be easier to add up only the 0, 5, 10, 18, and 12 (the numbers in parentheses)?

The use of the frequency distribution allows for this easier method. The extensions (xf) in Table 2-7 replace the sum of all the repeated values (one at a time). Notice that in Table 2-7 we are adding up the numbers shown previously

Table 2-7

x	f	xf
0	1	0
1	5	5
2	5	10
3	6	18
4	3	12
5	0	0
	20	45

as the sum of each block of numbers that were repeated. By adding another column to our frequency distribution table, we will obtain another extension that will enable us to do the same thing toward finding the sum of the x^2's (Σx^2) (Table 2-8).

Table 2-8

x	f	xf	$x^2 f$
0	1	0	0
1	5	5	5
2	5	10	20
3	6	18	54
4	3	12	48
5	0	0	0
	20	45	127

If we refer to Table 2-8, we will see that we can show subtotals of 0, 5, 20, 54, 48, and 0. The 20, for example, represents five values of $(2)^2$ being added to obtain x^2. The 54 represents six values of $(3)^2$, and so on. Thus by using the frequency distribution, we will find the values of Σf, Σxf, and $\Sigma x^2 f$, but they are actually replacements for n, Σx, and Σx^2, respectively. The f only indicates the use of the frequency distribution in its calculation. Therefore, the formula for mean ($\bar{x} = \Sigma x/n$) can become:

$$\bar{x} = \frac{\Sigma xf}{\Sigma f} \quad \text{or} \quad \frac{\Sigma xf}{n} \qquad \text{[formula 2-2], p. 50 in textbook]}$$

and the formula for the variance becomes:

$$s^2 = \frac{\Sigma x^2 f - \frac{(\Sigma xf)^2}{\Sigma f}}{(\Sigma f - 1)} \qquad \text{[formula (2-11), p. 64 in textbook]}$$

Remember: 1. Σf is the same as n.
2. Σxf is the same as Σx.
3. $\Sigma x^2 f$ is the same as Σx^2
The only difference is the method by which you obtain the three numbers.

Note: When the frequency distribution is used, there is no need to sum up the x column. It would make no sense at all to add up the various possible values without regard to the number of times that they occurred.

CHEBYSHEV'S THEOREM   Lesson 2-6

EXERCISES 2-5

1. Given the following set of data:

   1 1 1 2 2 2 2 3 3 3 3 3 4 4 4 4 4 5 5

   a. Calculate the sums Σx and Σx^2 using the individual numbers.
   b. Calculate the sums Σxf and $\Sigma x^2 f$ using the frequency distribution that represents this same set of data.
   c. Study the two sets of calculations and compare their similarities.

2. a. Given the following frequency distribution, find Σf, Σxf, and $\Sigma x^2 f$.

x	61	64	67	70	73
f	2	4	7	9	3

   b. Calculate $\bar{x}$ and s.

LESSON 2-6: CHEBYSHEV'S THEOREM

The theorem that is the subject of this lesson was developed by a Russian mathematician, P. L. Chebyshev (sometimes spelled Tchebysheff). You can think of this theorem as a sort of "minimum guarantee" for the proportion of data that will lie in a given interval about the mean of any distribution.

<u>Chebyshev's Theorem</u> (applies to any distribution of data): The proportion of any distribution that lies within k standard deviations of the mean is <u>at least $1 - 1/k^2$</u>, where k is any positive number larger than 1.

The number k is a number of multiples of the standard deviation. The standard deviation is a measure of distance, therefore "k standard deviations" becomes a distance also. For example, suppose we wanted k to equal 2 in a situation where the standard deviation was the value 5.2 units. k standard deviations would then be 2 multiples of 5.2, or 10.4 units. Suppose the mean of this situation is 30. To be within k standard deviations means to be somewhere between the two values, one smaller and one larger, that are exactly k standard deviations from the mean (Figure 3-1). The smaller value would be 19.6: 30 - 2(5.2) = <u>19.6</u>. The larger value would be 40.4: 30 + 10.4 = <u>40.4</u>.

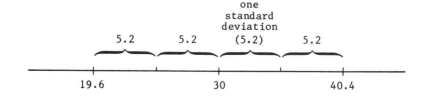

Figure 2-1

48 DESCRIPTIVE ANALYSIS: SINGLE-VARIABLE DATA   Chapter 2

Chebyshev's theorem says that if we inspect the interval from 19.6 to 40.4, we will find at least $1 - 1/k^2$ of the data. In this case we are using $k = 2$. $1 - 1/k^2 = 1 - 1/(2)^2 = 1 - 1/4 = 3/4$; therefore when we inspect the data for this situation we can be sure of finding "at least three-fourths" of the data somewhere between 19.6 and 40.4.

Special emphasis should be placed on the "at least" phrase, since we will find something in excess of 75 percent of the data in this interval. Remember our comment that this theorem was like a minimum guarantee — you will find *that much or more.* And it will almost always be *more:* (We do not remember ever seeing a set of data that did not exceed Chebyshev's proportion. We suppose that a "made-up" set could be produced, but it is rather doubtful that experimental data will do anything other than exceed this minimal value.) Chebyshev's theorem will most often be used as a mental check to see if the calculated values are realistic. The theorem should also give you some feeling for the measure "one standard deviation."

Chebyshev's theorem typically employs values of k that are whole numbers — $k = 2, 3,$ or 4 are about all that are ever used (2 and 3 being the most popular). When these values are considered for k, the proportions that are guaranteed are shown in Table 2-9.

Table 2-9

k	$1 - 1/k^2$
2	$1 - 1/(2)^2 = 1 - 1/4 = 3/4 = 75\%$
3	$1 - 1/(3)^2 = 1 - 1/9 = 8/9 = 89\%$
4	$1 - 1/(4)^2 = 1 - 1/16 = 15/16 = 94\%$

Again we would remind you that these are minimal values and you can expect to find a larger proportion of data. If you are interested in using a fractional value for k, say $k = 1\ 1/2\ (3/2)$, that is fine. $1 - 1/(3/2)^2 = 1 - 1/(9/4) = 1 - 4/9 = 5/9 = .55 = 55\%$. Thus, within 1 1/2 standard deviations of the mean we are sure to find more than 55 percent of the data.

You might wonder at this point, how do we find what proportion of the data is present in such an interval? Let's look at an illustration and find out.

Illustration Use the distribution of data presented in Lesson 2-5 (0, 1, 1, 1, 1, 1, 2, 2, 2, 2, 2, 3, 3, 3, 3, 3, 3, 4, 4, 4) and find the proportion of this data that falls within (a) 1.5 and (b) 2 standard deviations of the mean.

Solution First let's draw a dot-array diagram so that we can pictorially see our data (Figure 2-2).

CHEBYSHEV'S THEOREM   Lesson 2-6

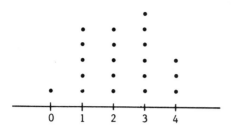

Figure 2-2

The mean of this data is 2.25 (45/20 = 2.25) and the standard deviation is 1.16 (as founed in Lesson 2-2).

a. $k = 1.5$ (or 3/2): Of the two values that are exactly 1.5 standard deviations from the mean, the smaller one is 2.25 - (1.5)(1.16) = 2.25 - 1.74 = 0.51, and the larger one is 2.25 + (1.5)(1.16) = 2.25 + 1.74 = 3.99. Thus we must find the proportion of data that falls inside the interval from 0.51 to 3.99 (Figure 2-3).

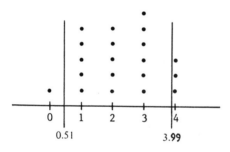

Figure 2-3

Simply count the values that are between 0.51 and 3.99. There are 16 — that is, 16 of the total 20 pieces of data. Thus 16/20 or 80 percent of our data is within 1.5 standard deviations of the mean. We expected to find at least 55 percent (in excess of 55 percent) and we did. Therefore we can conclude that the information we found does agree with the theorem.

b. Using $k = 2$, we find 2.25 - 2(1.16) = 2.25 - 2.32 = -0.07, and 2.25 + 2(1.16) = 2.25 + 2.32 = 4.57 (Figure 2-4). The interval from -0.07 to 4.57 includes all of our data (100 percent of it.)

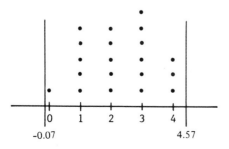

Figure 2-4

100 percent exceeds the 75 percent that was guaranteed. Therefore our findings again agree with the statement of the theorem.

If the data are given in the form of a grouped frequency distribution, we will not be able to find the number of data that belong to the interval simply by counting as in Illustration 1. Illustration 2 shows the adjustment that should be made for this situation.

Illustration Using the distribution of grouped data in Table 2-10 where the mean is 73.95 and standard deviation is 12.68, find the proportion of this data that is within 1.5 deviations of the mean.

Table 2-10

Class limits	f
40–49	2
50–59	5
60–69	12
70–79	17
80–89	13
90–99	6
	55

Solution The mean and standard deviation are used to obtain the values at the ends of the interval in question.

$$73.95 - (1.5)(12.68) = 54.93 = 54.9$$

$$73.95 + (1.5)(12.68) = 92.97 = 93.0$$

Let's place these two values on a histogram (bar graph) of the distribution so that we may better visualize what is happening (Figure 2-5).

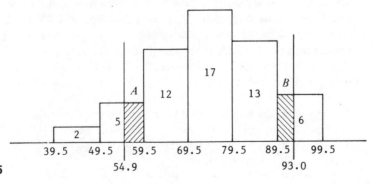

Figure 2-5

THE EMPIRICAL RULE Lesson 2-7

All the classes from 59.5 to 89.5 completely belong to our interval (54.9 to 93.0), thus we know that there are 42 (12 + 17 + 13) elements within our interval. However, the two shaded parts of the histogram represent a certain number of elements that also belong to the interval 54.9 to 93.0. So how many elements belong to each section? Section A represents 46/100 of that bar on the histogram [46/100 because its width is 4.6 (59.5 - 54.9) as compared to the width of the class, 10 (59.5 - 49.5)]. If we assume that the 5 items belonging to this class are "evenly distributed" over the class, we then will find 46/100 of the 5 in the shaded area. Therefore shaded area A represents 2.3 (46/100 of 5) pieces of data. Now looking at area B, we see that 35/100 of the bar containing 6 elements is within the interval (93.0 - 89.5 = 3.5; 99.5 - 89.5 = 10; and 3.5/10 = 35/100). 35/100 of 6 is 2.1. Therefore shaded area B represents 2.1 elements. Thus the interval from 54.9 to 93.0 represents a total of 46.4 of the 55 elements in the set of data. The proportion found is 46.4/55 = .843 or 84 percent.

The assumption that the data within a class are evenly distributed is necessary if the question is to be answered. Using the raw data is easier and more accurate; however, if the original values are "lost," then this interpolation technique is reasonably accurate.

EXERCISE 2-6

1. Use the frequency distribution found in Exercise 1 of Lesson 2-1.
   a. Find the mean and the standard deviation.
   b. Determine the proportion of data that is within 2 standard deviations of the mean and compare this with Chebyshev's theorem.

LESSON 2-7: THE EMPIRICAL RULE

The empirical rule is statement of fact about a "normal distribution." Later in the course we will discuss in more detail the so-called normal curve, but for now the empirical rule will give us some insight into this familiar distribution. The normal distribution possesses these properties: (1) it is symmetrically divided by its mean; (2) it is mounded at its mean; and (3) it displays the proportions stated in the empirical rule, which follows.

The Empirical Rule In a normal distribution, 68 percent of the data is within one standard deviation of the mean, 95 percent of the data is within two standard deviations of the mean, and 99.7 percent of the data is within three standard deviations of the mean.

If a distribution displays these three proportions for the three intervals about the mean, then it can be said to be approximately normal.

The term "normal" is an identifying name given to a particular distribution of data. The normal distribution is by far the single most important distribution, since it is the one that occurs most frequently in data. Normal, however, does not mean typical — some variables are typically distributed in a pattern quite different from the "normal" one.

The empirical rule can be used in one of two ways. (1) In a given distribution, we can calculate the proportion of data that lies within each of the three intervals. If the proportions found are close* to the three values in the empirical rule (68 percent, 95 percent, 99.7 percent), then we may conclude that the data are approximately normally distributed. (2) If we are told that a distribution is normal and wish to check the validity of the claim, we can find the three proportions and compare them to the percentages of 68, 95, and 99.7. If all three proportions are close to these values, then we must agree that the distribution does appear to be normal. Otherwise we will disagree with the claim.

To calculate the proportion of data that actually falls within these intervals, follow the same procedure as shown in Lesson 2-6 for Chebyshev's theorem. Table 2-11 shows the proportions stated in the empirical rule.

Table 2-11

Interval	Empirical-rule proportion
$\bar{x} - s$ to $\bar{x} + s$	68%
$\bar{x} - 2s$ to $\bar{x} + 2s$	95%
$\bar{x} - 3s$ to $\bar{x} + 3s$	99.7%

There are no supplementary exercises for this lesson. (However, a review of Lesson 2-6 will most likely answer any question you might still have about finding the proportion within a certain interval.)

SELF-CORRECTING EXERCISES FOR CHAPTER 2

1. Consider the sample 6, 4, 5, 9. Find each of the following.
   a. n b. Σx c. its mean, $\bar{x}$

2. Consider the sample 3, 5, 1, 3, 2, 7. Find each of the following.
   a. n b. Σx c. its mean, $\bar{x}$

3. A sample of eight students who had completed a take-home final exam for a math course reported that they had each spent the following number of hours completing the exam: 9, 6, 8, 9, 11, 7, 9, 13.

SELF-CORRECTING EXERCISES

   a. What is the response variable?
   b. Find n.
   c. Find Σx.
   d. Find the sample mean, $\bar{x}$.

4. A random sample of students was polled and asked, "How many hours did you spend studying during the weekend?" The data: 4, 8, 12, 3, 2, 7, 6, 9, 9, 13, 6, 5.
   a. What is the response variable?
   b. What type of variable is it? (attribute, continuous, discrete)
   c. Find n.
   d. Find Σx.
   e. Find the sample mean.

5. Find the variance s^2 for the data given in Exercise 1. [Use formula 2-7).]

6. Find the variance s^2 for the data given in Exercise 2. [Use formula 2-7).]

7. Using the sample data given in Exercise 2, find each of the following.
   a. the variance, using formula (2-7)
   b. the variance, using formula (2-10)
   c. the standard deviation

8. Using the sample data given in Exercise 4, find each of the following.
   a. the variance, using formula (2-7)
   b. the variance, using formula (2-10)
   c. the standard deviation

9. A mail order house is offering a $2.00 rebate on several hundred items. A random sample of seven items was taken, with the following regular prices (to the nearest dollar) being recorded for each: 10, 13, 16, 17, 15, 20, 11.
   a. What is the response variable?
   b. Is this variable attribute, continuous, or discrete?
   c. Find the sample mean, $\bar{x}$.
   d. Find the sample standard deviation, s.

10. You are given the following set of twenty measurements:

    8 9 10 7 5 7 6 0 12 8
    3 10 11 4 9 16 4 -2 10 7

   a. Construct a dot-array diagram depicting this set of data.
   b. Estimate $\bar{x}$ and s.
   c. Calculate $\bar{x}$ and s.
   d. Compare your calculated answers in part c to your estimates in part b.

11. The following set of body temperatures (rounded to the nearest degree) was taken at a local clinic:

    100 100 101 101 102 103 104 105

    Calculate the following:
    a. the mean
    b. the standard deviation

12. Given the same set of body temperatures (not rounded off):

    100.0 100.4 101.0 101.4 101.8 103.2 104.4 105.2

    Calculate the following:
    a. the mean
    b. the standard deviation
    c. Comment on the effect of rounding off the data. Do you think this is the typical effect of rounding off? Explain.

13. The following ungrouped frequency distribution represents the number of children as reported by 90 fathers at a PTA meeting.

x	f
1	9
2	18
3	21
4	16
5	12
6	5
7	3
8	3
9	3

    a. Construct a frequency histogram depicting this data.
    b. Construct a relative frequency histogram depicting this same data.

14. Use the block of 50 single-digit random numbers from the upper left-hand corner of the random-number table on page 566 of your textbook as your sample.
    a. Construct an ungrouped frequency distribution of the sample.
    b. Construct a stem-and-leaf diagram depicting these 50 random digits.
    c. Draw a frequency histogram showing this sample.
    d. Form the cumulative frequency distribution corresponding to the distribution asked for in part a.
    e. Draw an ogive (a cumulative relative frequency polygon) of this same sample.

SELF-CORRECTING EXERCISES

15. The following set of data represents the ages, in years, of the oldest brother or sister of 80 students who attended the lecture given last week by a visiting celebrity.

Ages of oldest brother or sister

25	17	22	27	22	14	15	26
23	26	25	23	24	25	27	31
10	22	16	21	30	17	39	26
24	31	26	35	29	20	16	24
30	11	20	17	43	16	22	14
25	15	22	25	20	15	22	10
42	33	22	25	22	20	29	25
28	21	21	26	17	20	26	17
14	16	20	27	25	25	22	23
19	20	22	27	31	19	23	36

a. Construct a grouped frequency distribution of this data using 9–13, 14–18, ..., 39–43 as your class limits.
b. Draw a relative frequency histogram of this data using class marks to label the horizontal scale.
c. Construct a cumulative grouped frequency distribution of this same set of data.
d. Draw an ogive showing the above sample.

16. The following set of data represents the weight of luggage for 50 passengers who were flying overseas.

Luggage weights

35.1	51.4	42.6	47.2	49.7
37.1	49.2	45.8	41.8	50.1
40.7	43.9	41.5	50.9	43.1
47.8	47.6	34.5	45.7	47.6
46.2	45.3	48.5	45.6	40.8
44.4	28.9	50.1	44.3	41.6
38.4	38.3	40.3	39.7	39.9
39.1	40.2	42.5	46.8	43.7
33.7	38.5	38.6	39.9	46.8
37.3	38.9	39.8	42.5	49.2

a. Construct a group frequency distribution of this data using 28.0–30.9, 31.0–33.9, ..., 49.0–51.9 as the class limits.
b. Draw a histogram showing this data.
c. Draw an ogive showing this data.

17. The following grouped frequency distribution represents a sample of 150 counts taken by a geiger counter. (x is the number of particles counted per unit of time.)

Class limits	Frequency
21–25	3
26–30	11
31–35	19
36–40	28
41–45	24
46–50	23
51–55	19
56–60	15
61–65	8
	150

a. Draw a frequency histogram of this data. Use class marks to label the horizontal scale.
b. Draw an ogive of this sample data.

18. Consider the following set of 40 pieces of data.

2, 3, 3, 5, 6, 6, 7, 7, 7, 8, 8, 9, 9, 9, 9, 9, 10, 10, 10, 10, 11, 11, 11,

11, 12, 12, 12, 12, 12, 12, 13, 13, 14, 14, 15, 15, 15, 16, 17, 19

a. Form an ungrouped frequency distribution.
b. Construct a stem-and-leaf diagram depicting the above data.
c. Draw a histogram showing this distribution.
d. Find the summations Σxf and $\Sigma x^2 f$.
e. Calculate the mean $\bar{x}$.
f. Calculate the standard deviation s.
g. Form a grouped frequency distribution of the above data. (Use the class structure that puts 2, 3, and 4 all in the first class.)
h. Draw a histogram of this grouped distribution.
i. Find the summations Σxf and $\Sigma x^2 f$ as applied to this grouped distribution.
j. Calculate $\bar{x}$.
k. Calculate s.
l. Compare the answers to parts i, j, and k, to the answers of parts d, e, and f. Why are these values similar but different?

SELF-CORRECTING EXERCISES

19. Using the data and the ungrouped distribution from Exercise 13, find the following.
    a. the totals Σxf and $\Sigma x^2 f$
    b. the mean number of children
    c. the standard deviation for the number of children

20. Using the ungrouped frequency distribution of random digits from Exercise 2, find the following.
    a. the summations Σxf and $\Sigma x^2 f$
    b. the mean value $\bar{x}$
    c. the standard deviation s

21. Using the grouped frequency distribution from Exercise 15, find the following.
    a. the summations Σxf and $\Sigma x^2 f$
    b. the mean age of the oldest brother or sister
    c. the standard deviation of their ages

22. Using the grouped frequency distribution from Exercise 16, find the mean and standard deviation of the luggage weights.

23. Calculate the mean and the standard deviation of the geiger counter counts presented in the grouped frequency distribution of Exercise 17.

24. For the data presented in the ungrouped frequency distribution in Exercise 13, find the following.
    a. the median
    b. the mode
    c. the first quartile
    d. the 30th percentile
    e. the range

25. For the luggage weight data in Exercise 16, find the following.
    a. the median
    b. the third quartile
    c. the 45th percentile

26. For the ages of the oldest brothers or sisters in Exercise 15, find the following.
    a. the median
    b. the midquartile
    c. the midrange
    d. the range

DESCRIPTIVE ANALYSIS AND PRESENTATION OF BIVARIATE DATA

3

LESSON 3-1: THE CORRELATION COEFFICIENT

Correlation analysis is performed on bivariate data (two pieces of data that differ from each other but are related in some way) to measure the strength of the linear relationship between the two variables. In analyzing the strength of this relationship, we would ask the following question: As the x variable increases in value, does the y variable (1) increase in value, (2) remain at about the same value, or (3) decrease in value? The answer can usually be found by inspecting a scatter diagram of the data, which pictorially represents the pattern of behavior of the two variables.

To measure the predictability of this pattern mathematically, we will calculate a single number, which will give us a "standardized" measure. This number is the product of the correlation analysis; it is commonly called the linear correlation coefficient and is symbolized by r.

Note: The adjective "standardized" means that the measure of correlation obtained has been adjusted so that the amount of spread within the data does not affect the resulting value. A measure that depended partly upon the amount of spread within the data would be of no use in making comparisons. A further discussion of this standardized measure will be found in Chapter 13.

The linear correlation coefficient r is a numerical value which must take on a value somewhere between -1 and $+1$. Remember that we want to measure (1) how y tends to behave as x increases (does y increase, remain the same, or decrease?) and (2) how close this pattern of change is to a straight line. The direction of change in y will be indicated by the sign of r. r will be positive if y tends to increase as x increases; r will be negative if y tends to decrease as x increases; r will be zero if y does not change as x increases; and r will be close to zero in value if there is little or no consistent linear pattern of change. If the set of data does display a linear pattern, the numerical value of r (the number itself without the accompanying sign) will be close to 1. Figure 3-1 shows the real number line from -1 to $+1$.

CORRELATION COEFFICIENT   Lesson 3-1

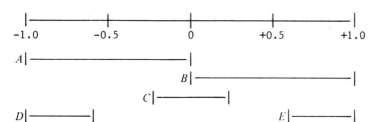

Figure 3-1

The line segments A, B, C, D, and E represent various values of r. If the value calculated for r belonged to the A interval (negative value), we would have mathematical evidence that the value of y decreases as x increases (that is, the scatter diagram would be in a "downhill" position; see page 109 in the textbook). The values of r that fall within interval B are positive, meaning that the value of y tends to increase as x increases. (The scatter diagram will show an "uphill" position.) Line segment C represents those values of r that are near zero. The linear correlation coefficient will be near zero when the pattern is nearly horizontal (no correlation) or when the pattern is not linear. Values of r belonging to interval D suggest a strong linear relationship where y is decreasing as x increases. E represents values of r that show a strong linear relationship with y increasing as x increases.

EXERCISES 3-1

1. Determine whether the following pairs of variables would show a positive or a negative value for r.
   a. children's age and height (newborn to junior-high-school age)
   b. a serviceman's base salary and his rank
   c. a subject's intelligence and the length of time required to solve an experimental problem
   d. the age of an automobile and the annual cost of repair
   e. the age of an automobile (years) and its trade-in value (dollars)
   f. a student's midterm average and his final grade in this course

LESSON 3-2: THE LINE OF BEST FIT

The "line of best fit" is the line that approximates the relationship between two variables. In our study it will be the straight line that best approximates the path followed by the data as shown on the scatter diagram. The true linear relationship, that is, the equation we would calculate if we had the entire population of data available, is expressed by the equation $\hat{y} = \beta_0 + \beta_1 x$, where β_0 is the y-intercept and β_1 is the slope of this straight line. (See Lesson I-4 in this study guide.) The equation of the line of best fit (straight line of best fit) is expressed by $\hat{y} = b_0 + b_1 x$ for the sample data. b_0 and b_1 are the sample statistics that are used to approximate the true values of the population parameters, β_0 and β_1, respectively. The line of best fit is found when the sum of the squares of all the lengths of line segments shown in Figure 3-16 (page 115 in the textbook) is as small as possible. Each point (x,y) of data lies a certain vertical distance from any potential line of best fit. The point on the line of best fit that is directly above or below (x,y) is called $(x,\hat{y})$. The vertical distance between these two points is $(y - \hat{y})$. The y-intercept (b_0) and slope (b_1) of the line of best fit must be values that cause the sum of the squares of $(y - \hat{y})$, $\Sigma(y - \hat{y})^2$, to be as small as possible. The values of b_0 and b_1 (the constants satisfying the least-squares criterion) are calculated by using formulas (3-6) and (3-7), page 115 of the textbook.

It should be pointed out that the equational relationship between x and y, where x is the independent variable (input variable, horizontal axis), is different from the equational relationship where y is the independent variable. In some situations you can choose which variable is to be the input variable, but once you make this choice and perform the calculations for b_0 and b_1, you do *not* have the liberty of using the other variable as the input variable. Interchanging the roles of input and output variables in effect changes the "best-fit" criterion from vertical line segments to horizontal segments. Figure 3-16 on page 99 in the textbook shows x as input and y as output. When the roles of x and y are interchanged, the values whose sum of squares is now to be minimized are shown in Figure 3-2, next page. The least-squares criterion used here would cause a different equational relationship.

EXERCISES 3-2

1. a. What is the "least-squares" criterion used to define the line of best fit?
   b. What would this "least" possible value be and when would it occur?

MAKING PREDICTIONS   Lesson 3-3

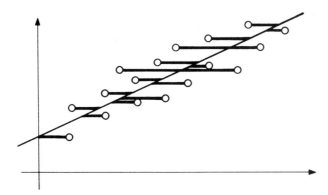

Figure 3-2

2. Why does the interchanging of roles of the two variables cause a different equation to result as the line of best fit?

3. What would be the interpretation of a situation where the slope (b_1) of the line of best fit was found to be zero? negative? positive?

LESSON 3-3: MAKING PREDICTIONS

To make a prediction we need only the equation and the value of the input variable that the predicted value is to be based on. Let's use Illustration 3-6 on page 117 in the textbook. The calculated equation of the line of best fit is $\hat{y} = -186.5 + 4.7x$, where x and y represent the height and weight, respectively, of college women. A person from this population is 64 inches tall and you wish to predict her weight by means of the line of best fit. By substituting 64 in place of x in the equation, we will obtain a value for $\hat{y}$. ($\hat{y} = -186.5 + 4.7(64) = -186.5 + 300.8 = 114.3$) $\hat{y}$ is the predicted value for y and is 114.3 pounds.

Think about this prediction for a minute. What does it mean? If another woman says she is 64 inches tall, do you also expect her to weigh 114.3 pounds? (In fact, within the population there are several persons whose height is 64 inches.) The 114.3 pounds would more likely be thought of as an approximate mean value for all persons who are 64 inches tall. It would then make sense to predict a 64-inch-tall woman's weight individually as the mean value. Actually we would expect very few to weigh exactly 114.3, but perhaps many would weigh "close" to it. Further discussion along this line must wait until some other ideas have been developed. We might conclude at this point that we would expect our prediction to average out as being reasonably accurate; each individual prediction would be off by a small amount, but we would expect these amounts to balance out.

EXERCISES 3-3

1. What is the main use of the line of best fit?

2. a. What does it mean to predict a value of y based on a known value of x?
   b. What is the interpretation of the calculated value, $\hat{y}$?

3. $\hat{y} = 10 + 2.5x$ is given as the line of best fit. What value of y can be expected in a situation where x is equal to 3? 8?

SELF-CORRECTING EXERCISES FOR CHAPTER 3

1. Inspect each of the following scatter diagrams and determine whether or not the set of data appears to be linear in nature.

   a.
   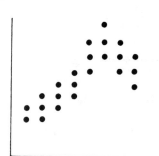

   b.

   c.
   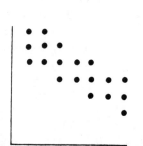

   d.
   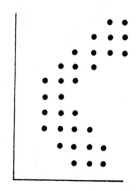

SELF-CORRECTING EXERCISES

2. Plot a scatter diagram of each of the following sets of ordered pairs on separate axis systems.

   a.

x	1	2	3	4	4	5
y	1	1	2	2	3	3

   b.

x	1	2	3	4	5	6	7	8	9	10
y	5	4	4	3	2	2	1	0	0	0

   c.

x	0	2	3	1	6	2	1	3	5	4	2	5
y	45	20	10	35	5	15	30	15	10	10	25	5

3. Identify by letter the line segment or point that contains your estimate for the linear correlation coefficient for each of the scatter diagrams shown below.

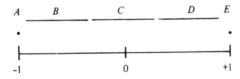

a.

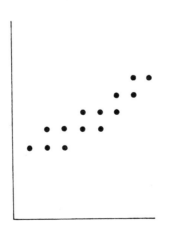

b.

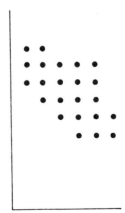

c.

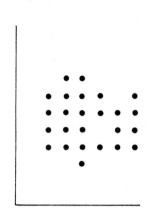

d.

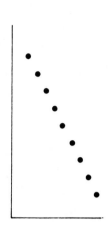

e.

f.

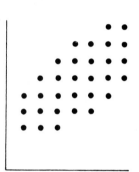

4. If the following variables were to be paired and a correlation analysis carried out, would you expect the calculated value of r to be positive, negative, or near zero?
   a. a boy's age and his shoe size
   b. the liveliness of a golf ball and its age
   c. a person's age and the age of his automobile
   d. a person's salary and his length of service to his company
   e. the distance that a college student commutes to college and his grade-point average

5.

x	0	1	2	3	4	5
y	1	2	2	3	3	4

a. Plot a scatter diagram of the above set of paired data (x,y).

SELF-CORRECTING EXERCISES

b. Construct the necessary chart and perform the extensions to obtain the totals Σx, Σy, Σx^2, Σxy, Σy^2.
c. Using formula (3-2), page 106 in the textbook, calculate the linear correlation coefficient, r.
d. Using formulas (3-6) and (3-7), page 115 in the textbook, calculate the two coefficients, b_0 and b_1.
e. Write the equation of the line of best fit.

6.

x	0	1	1	2	3	3	4	5	5	6	7	7	8
y	4	4	3	3	3	2	2	2	1	1	1	0	0

a. Plot a scatter diagram of the above set of paired data (x,y).
b. Construct the necessary chart and perform the extensions to obtain the totals Σx, Σy, Σx^2, Σxy, Σy^2.
c. Using formula (3-2), page 106 in the textbook, calculate the linear correlation coefficient r.
d. Using formulas (3-6) and (3-7), page 115 in the textbook, calculate the two coefficients, b_0 and b_1.
e. Write the equation of the line of best fit.

7. A random sample of 16 students was obtained from a teacher's class roster for last semester. A random sample of students' pre-final-exam class averages (x) and the students' scores on the final examination (y) were recorded.

Student	1	2	3	4	5	6	7	8	9	10	11	12	13	14	15	16
x	76	61	74	94	79	63	80	87	91	50	83	85	66	72	53	90
y	68	55	77	99	81	64	78	88	86	57	93	82	64	67	50	89

a. Plot these ordered pairs (x,y) on a scatter diagram.
b. Calculate the extensions x^2, xy, and y^2, and find the summations Σx, Σy, Σx^2, Σxy, and Σy^2.
c. Calculate the linear correlation coefficient r.
d. Calculate the coefficients b_0 and b_1 and write the equation of the line of best fit.

8. The body temperature of patients may be taken in different ways, two of which are the oral and rectal methods. The following set of data represents a sample taken to check the validity of the statement: "The rectal temperature is one degree higher than the oral temperature."

Patient	1	2	3	4	5	6
Oral (x)	98.2	100.4	101.8	94.4	98.6	103.0
Rect. (y)	99.2	101.2	102.0	95.6	99.8	104.0

DESCRIPTIVE ANALYSIS: BIVARIATE DATA   Chapter 3

Patient	7	8	9	10	11	12
Oral (x)	101.0	96.8	105.2	101.4	95.2	100.0
Rect. (y)	102.0	98.0	106.2	102.6	96.4	101.0

a. Plot this set of data on a scatter diagram.
b. Calculate the extensions x^2, xy, and y^2, and the totals Σx, Σy, Σx^2, Σxy, and Σy^2.
c. Calculate the linear correlation coefficient r.
d. Calculate the coefficients b_0 and b_1 and write the equation of the line of best fit.
e. Based on the results obtained in the answers to parts a through d, comment on what this evidence indicates about the statement mentioned above.

9. Using the equation for the line of best fit found in Exercise 8d, find the expected rectal temperature for a patient whose oral temperature is:
   a. 100.0
   b. 98.0
   c. 102.0
   d. 98.6
   e. 103.4
   f. Compare the answers for parts a through e with the value claimed by the statement: "Rectal temperature is one degree higher than oral temperature."

10. Using the equation for the line of best fit found in Exercise 7d, find the final exam score that would be predicted for the student with a pre-final average of:
    a. 60
    b. 80
    c. 92

PROBABILITY

4

LESSON 4-1: MUTUALLY EXCLUSIVE AND ALL-INCLUSIVE EVENTS

When dealing with a probability problem, one must first perceive the sample space, that is, it must be either completely "listed" or visualized mentally. ("Listing" is much more appropriate for the beginner.) Recall that a sample space is the set of "possibilities" (the set of all possible results that might occur) expressed as simple events. Any one event may be defined as desired, but it will be composed of one or more of the sample points (simple events) from the **sample space.**

Illustration   A box contains eight tags with one name on each: Ann, Betsy, Connie, Debbie, Ed, Frank, George, and Henry. The probability experiment is to draw one name. The sample space represents the list of possible names that could be drawn — the same list as above. The eight names represent the eight different individual results that could occur when one name is drawn from the box.

Let's define a set of events. (1) Event P is "a girl's name is drawn" (this event occurs if the one name drawn happens to be a girl's name — Ann, Betsy, Connie, Debbie). (2) Event Q is "the name drawn is spelled with 5 letters" (event Q occurs if Betsy, Frank, or Henry is the name drawn). (3) And event R is "the name drawn is spelled with less than 5 letters" (event R occurs if Ann or Ed's name is drawn). Each of these events is made up of more than one sample point.

$$\text{Event } P = \{\text{Ann, Betsy, Connie, Debbie}\}$$
$$\text{Event } Q = \{\text{Betsy, Frank, Henry}\}$$
$$\text{Event } R = \{\text{Ann, Ed}\}$$

Remember: An event is said to occur if any one of the elements listed within the definition of its set occurs.

Mutually exclusive events.   A set of events is mutually exclusive if the occurrence of any one of these events excludes the occurrence of all the other events within that set. (Any one of these events may happen, but only one at a time.)

The above set of events (*P, Q,* and *R*) as a group does not display this mutually exclusive property. If "Ann" happens to be selected, then both events *P* and *R* have occurred; if "Betsy" is selected, then both *P* and *Q* occurred. If we were to consider only events *Q* and *R*, however, we would see that they are a mutually exclusive pair — the definitions of these two events make it impossible for both to occur as the result of one drawing. The two sets that make up these events are disjoint.

All-inclusive events. A set of events is all-inclusive if the events are defined in such a way that at least one of the events occurs when each element of the sample space results, in other words, the list of event sets completely covers the entire sample space.

Let's add a fourth event, *S,* to our previous set. Let it be "the name drawn is spelled with more than 5 letters."

Event *S* = {Connie, Debbie, George}

Our events, *P, Q, R,* and *S,* form a set of events which are all-inclusive. Consider each element of the sample space: (1) If Ann is selected, then events *P* and *R* occur; (2) if Betsy is selected, then events *P* and *Q* occur; (3) if Connie, then *P* and *S*; (4) if Debbie, then *P* and *S*; (5) if Ed, then *R*; (6) if Frank, then *Q*; (7) if George, then *S*; (8) and if Henry, then *Q*. You see, no matter which element was selected, one or more of the events occurred. Thus, events *P, Q, R,* and *S* are all-inclusive events.

Let's consider the set of events *Q, R,* and *S* as defined above on the sample space of eight names. This set of events displays both the mutually exclusive and the all-inclusive properties. Let's approach this in two ways: (1) look at the list of the sample space and the lists for each individual event, and (2) look at their definitions. (Use the first letter to identify each name.)

1. Sample space: {A, B, C, D, E, F, G, H}

   Event *Q*: {B, F, H}
   Event *R*: {A, E}
   Event *S*: {C, D, G}

Notice that the set of possibilities corresponding to the events *Q, R,* and *S* is disjoint; that is, there is no one element that belongs to more than exactly one of these sets. Therefore *Q, R,* and *S* are mutually exclusive. Further, if you check each element of the sample space you will notice that it belongs to an event set. Thus these events are all-inclusive.

2. Consider the definitions:

   Q: exactly 5 letters
   R: less than 5 letters
   S: more than 5 letters

It seems quite clear that one of these events must always occur (what else is

COMPOUND EVENTS   Lesson 4-2

there?). This then demonstrates the all-inclusive property. It also seems quite clear that only one of these events could occur at a time. (How could a name be spelled with less than 5 letters and at the same time be spelled with exactly five letters? And so on.) This then demonstrates the mutually exclusive property.

EXERCISES 4-1

1. True or False:
   a. Any event and its complement are mutually exclusive events.
   b. Any event and its complement are all-inclusive events.

2. Consider the sample space of single-digit integers: {0, 1, 2, 3, . . . , 9}, and let the following events be defined as shown:

   Event A: {2, 4, 6, 8}
   Event B: {1, 3, 5, 7, 9}
   Event C: {0, 1, 2, 3}

   a. Are events A and B mutually exclusive? Explain.
   b. Are events A and B all-inclusive? Explain.
   c. Are events A and C mutually exclusive? Explain.
   d. Are events A, B, and C all-inclusive? Explain.
   e. Define an event D such that the set of events A, B, and D will display both the mutually exclusive and the all-inclusive properties.
   f. Define an event E such that events C and E are mutually exclusive and all-inclusive.

LESSON 4-2: PROBABILITIES OF COMPOUND EVENTS USING RELATIVE FREQUENCY

The probability that any event occurs is defined as the relative frequency with which that event occurs or can be expected to occur. A relative frequency is nothing more than a ratio (fraction) which compares the size of one number to the size of a second number. The ratio of 4 to 8 can be expressed as 4/8 or 1/2. If something happens 4 times out of 8 possible chances, then its relative frequency is 4/8 (four-eighths) or 1/2.

A compound event is any event which is composed of two or more of the simple events which make up the sample space for a problem. The probability for any event may be found by determining the solution set for that event, counting the number of elements belonging to the solution set, and then dividing this count by the number of elements in the sample space. This method will work for all probabilities where a sample space of equally likely possibilities can be found.

Illustration   Let's use Experiment 4-6 on page 130 in the textbook, the rolling of two dice. The sample space is shown in Table 4-1 (below) as an array of ordered pairs. The first digit represents the number of dots showing on the white die, the second digit is the number of dots showing on the black die.

Table 4-1

16	26	36	46	56	66
15	25	35	45	55	65
14	24	34	44	54	64
13	23	33	43	53	63
12	22	32	42	52	62
11	21	31	41	51	61

$n(S) = 36$

Define event N to be "a sum of nine." Event N is a compound event which occurs if any one of four different simple events occurs: N = {36,45,54,63}. $n(N) = 4$. Therefore N occurs on 4 of the 36 sample points and P(sum of 9) = 4/36 = 1/9.

Define event F as "the white die is 4." Event F will occur when one of the following occurs: {41, 42, 43, 44, 45, 46}. Therefore $n(F) = 6$ and $P(F) = 6/36$ = 1/6.

Basically the same technique can be applied to compound events formed by the connectives "and" and "or." Using the above events, N and F, find the probability of (F and N) and (F or N). The event (F and N) will occur only when the sample point 45 occurs. The connective "and" means that both events, F and N, occurred on the particular throw of the dice. F occurs whenever 41, 42, 43, 44, 45, or 46 results. Event N occurs whenever 36, 45, 54, or 63 results. Only 45 is on both lists; therefore, (F and N) = {45}, and $n(F$ and $N) = 1$; so $P(F$ and $N) = 1/36$. The connective "or" is interpreted to mean "one event occurs, or the other event occurs, or both events occur." Thus (F or N) = {41, 42, 43, 44, 45, 46, 36, 54, 63}. (Notice that 45 is listed only once.) $n(F$ or $N) = 9$, and therefore $P(F$ or $N) = 9/36 = 1/4$.

Note: In the text we have differentiated between theoretical (expected) probabilities [$P(\ )$] and experimental (observed) probabilities [$P'(\ )$] by use of the prime sign ('). The observed probability is always the relative frequency with which an event occurs [$P'(A)$ = number of $(A)/n$]. However, the relative frequency concept for the theoretical probability can only be used in connection with a sample space of equally likely events.

Let's take another look at compound events using "or." The connective "or" is associated with the set operation "union." Consider the compound event (F or N) again. The solution set for event F is a set containing all the sample

71 **COMPOUND EVENTS**   **Lesson 4-2**

points for which the first digit is a 4. Think of these six elements as six objects in a container labeled *F* (Figure 4-1).

Figure 4-1

Now think of a container *N* (solution set *N*) that contains the four elements belonging to set *N* (Figure 4-2).

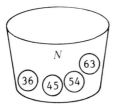

Figure 4-2

In effect, when the connective "or" is used (the union of two sets results), the solution set for the compound event is the set that results from "dumping" each of the individual sets into a "common container." This common container is the union, or the solution set, for the "or" compound event. Using (*F* or *N*) from above, Figure 4-3 shows us the result.

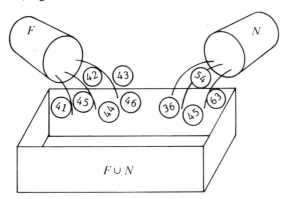

Figure 4-3

Thus $F \cup N$ now contains 41, 42, 43, 44, 45, 46, 36, 45, 54, and 63. However, there is only one element 45 in the sample space and thus only one element 45 in the union of sets *F* and *N*. The set $F \cup N$, therefore, contains exactly 9 elements. (Be extremely careful when working with "or" events.)

EXERCISES 4-2

1. Consider the probability experiment where one penny and one die are placed in a cup, shaken, and dumped out onto a table top. We are concerned with finding the probability of two compound events: $P(A$ and $B)$ and $P(A$ or $B)$, where A is the occurrence of a head on the penny and B is the occurrence of an even number of dots showing on the die.
   a. Find the sample space for this experiment. (There are 12 sample points.)
   b. List the sample points that belong to the solution set of $(A$ and $B)$. Hint: In order for a point to belong to event A, you must be able to answer yes to the question, Did a head result? Likewise, in order for a point to belong to set B, you must be able to answer yes to the queston, Did an even number of dots show? The answer must be yes to *both* questions if the element is to belong to $A \cap B$ (the solution set for A and B).
   c. Find $P(A$ and $B)$.
   d. List the sample points that belong to the solution set of $(A$ or $B)$, that is, $(A \cup B)$. Hint: For a point to belong to the union, you must be able to answer yes to one or both of the questions shown above for each individual sample point.
   e. Find $P(A$ or $B)$.

2. Using the experiment of rolling two dice, find the following probabilities.
   a. P (white die is even and black die is even)
   b. P (white die is even or black die is even)
   c. P (sum is 5 or black die is odd)
   d. P (sum is 5 and black die is odd)
   e. P (sum is 11 or black die is 3)
   f. P (sum is 11 and black die is 3)

LESSON 4-3: INDEPENDENT EVENTS

The term "independent" has much the same meaning in connection with probability as it has in other contexts. According to Webster's dictionary, independent means "not subject to bias or influence." In the realm of probability, the influence or lack of influence is shown in the value of the probability of an event. If an influence is shown (dependency), then the probability of an event will change as information is learned about another event. If the probability of an event is not caused to change owing to the additional knowledge about a second event, then the two events are said to be independent: there was "no effect" on the probability (it remained at the same value). To further demonstrate this, let's look at an illustration.

Illustration Let's consider the experiment of tossing a penny and a die (Exercise 1 of Lesson 4-2). Events A and B were defined "head" and "even number of dots," respectively. Are these two events independent?

INDEPENDENT EVENTS   Lesson 4-3

First, consider the entire sample space:

{H1, H2, H3, H4, H5, H6, T1, T2, T3, T4, T5, T6}

and determine the probability of each event separately. $P(A) = 6/12 = 1/2$ and $P(B) = 6/12 = 1/2$. (The two probabilities have the same values; however, this is of no consequence.) Now we are ready to consider the question, Are A and B independent?

Think of the experiment being performed in this manner. A friend places the penny and the die in a cup, shakes it, and dumps them into a box where you cannot see the result. If he gives you no "hint" as to what has occurred, you would have to say that the $P(A) = 1/2$ and $P(B) = 1/2$, as stated above. But suppose he tells you something about one of these events. Suppose he says, "Event B has occurred." With this information, determine the probability that event A has occurred. The sample space has now been reduced, since there are several points which you know did not occur.

Sample Space = {H̶1̶, H2, H̶3̶, H4, H̶5̶, H6, T̶1̶, T2, T̶3̶, T4, T̶5̶, T6}

The remaining list of possible occurrences now becomes the "reduced sample space" for the situation.

Reduced Sample Space = {H2, H4, H6, T2, T4, T6}

a set of six equally likely events. Now the probability that event A happened, knowing that event B has already occurred, $P(A,$ knowing $B)$, is 3/6 or 1/2. Thus the probability of A, $P(A)$, was not influenced by knowledge about event B. That is, $P(A) = P(A,$ knowing $B)$, and we must proclaim A and B to be independent events.

The "conditional knowledge" (the "hint" in the above illustration) can take almost any form. Something did or didn't happen, but the effect on the sample space will be to reduce it. If the remaining proportion of the other event is the same as it was in the whole sample space, then independence between event A and event B (the conditional knowledge) is established. If the proportions are different, then the events are dependent.

Let's consider the pair of dice experiment and the two events, F (4 on the white die) and N (sum of nine), as described in Lesson 4-2. Are these two events, F and N, independent? Again, the total sample space is our first concern (Table 4-2).

Table 4-2

16	26	36	46	56	66
15	25	35	45	55	65
14	24	34	44	54	64
13	23	33	43	53	63
12	22	32	42	52	62
11	21	31	41	51	61

The probability of F is $P(F) = 6/36 = 1/6$ and the probability of N is $P(N) = 4/36 = 1/9$. Again think of the dice being rolled, but out of your sight. Find the probability that N occurs having been told that F occurred.

If F occurred, the reduced sample space contains only six elements: {41, 42, 43, 44, 45, 46}, as the other 30 sample points have been ruled out. Now inspect this reduced sample space and find the probability of event N. $P(N,$ knowing $F)$ is 1/6. Previously (that is, without any clues), we found the probability of N to be 1/9; now it is 1/6. $P(N) \neq P(N,$ knowing $F)$. Thus it seems reasonable to say that the probability of event N is indeed influenced by knowledge about event F. Therefore, F and N are *not* independent events, but are dependent events.

There are several possible pairs of probabilities that may be compared to check for independence. Consider two events, A and B. If they are independent, then $P(A) = P(A,$ knowing $B) = P(A,$ knowing not $B)$, and $P(B) = P(B,$ knowing $A) = P(B,$ knowing not $A)$. (Any two of the first group or any two of the last group will serve as the check.)

A *conditional probability* is the probability of an event occurring given a conditional circumstance. $P(A,$ knowing B has already occurred) is a conditional probability. This particular conditional probability is also commonly referred to as $P(A,$ given $B)$, read, "the probability of A occurring given that B has already been observed." The notation used to express a conditional probability is $P(A|B)$. The event identified in front of the vertical line (A in this case) is the event whose probability is being sought. The event identified behind the vertical line (B in this case) is the "condition."

We will also be using the idea of independence in association with repeated trials. In this situation we will typically try to determine whether or not results (known or unknown) of one trial influence the results of any other trial. For example, let's consider an experiment which calls for tossing a single penny several times. The probability of heads on any one toss is 1/2. Suppose a head occurs on the first toss, what then is the probability of heads on the second toss? It is 1/2. The fact that heads occurred on any one toss does not influence the outcome of any other toss if the tossing is performed fairly. The individual trials (each toss) are then said to be independent trials.

EXERCISES 4-3

1. Consider the experiment of tossing a penny and a die, and events A and B as described earlier in this lesson. Find the following probabilities.
   a. $P(A,$ knowing B did not happen)
   b. $P(B,$ knowing $A)$
   c. $P(B,$ knowing $\bar{A})$

2. Consider the rolling of two dice (white, black) and events S (sum of seven), T (sum of ten), and F (four on the white die). Find the following.
   a. $P(S)$ and $P(S|T)$ [$P(S,$ knowing $T)$]

75   MULTIPLICATION RULE   Lesson 4-4

   b. Are events S and T independent? Explain.
   c. $P(S)$ and $P(S|F)$ [$P(S$, knowing $F)$]
   d. Are events S and F independent? Explain.
   e. $P(F)$ and $P(F|\bar{T})$ [$P(F$, knowing $\bar{T})$]
   f. Are F and T independent? Explain.
   g. $P(F)$ and $P(F|S)$ [$P(F$, knowing $S)$]
   h. Are F and S independent? Does this agree with the answers to parts c and d? Explain.

3. Consider the following two coin-tossing experiments.
   I. Toss a single coin and observe a head or a tail. Toss it a second and third time repeating the observations.
   II. Toss a single coin and observe a head or a tail. If a head occurred, stop. If a tail occurred, toss the coin again. Repeat; if second toss is a head, stop; if tail, then toss the coin for a third time.
   Consider the event "a head occurred," and determine whether or not the trials of the above experiments are independent.

4. Consider the act of drawing one playing card at random from an ordinary deck of 52 cards. Let F be the event that a face card is drawn. Let K be the event that a king is drawn. Find the following.
   a. $P(F)$
   b. $P(K)$
   c. $P(\bar{F})$
   d. $P(K|F)$ [$P(K$, knowing $F)$]
   e. $P(K|\bar{F})$ [$P(K$, knowing $\bar{F})$]
   f. Are events F and K independent? Explain.

LESSON 4-4: THE MULTIPLICATION RULE

In Lesson 4-2, we discussed ways to find the probabilities of compound events by using an equally likely sample space and the relative frequency definition as applied to a given event. The sample space concept is not always possible or efficient. In this lesson and in the next (Lesson 4-5) we should like to discuss two rules that allow us to calculate the probabilities of compound events by using the probabilities of the simple events involved. These rules will be somewhat restrictive, as we will not discuss all of the ramifications of the two rules.

The multiplication rule simply tells us to multiply the probabilities of the events which make up a compound event formed by the use of the connective "and." We will multiply the respective probabilities when the events involved are independent.

The Multiplication Rule (Case 1) If events A and B are independent events defined on the same sample space, then the probability that they both occur simultaneously is found by multiplying the probability of event A by the probability of event B.

$P(A \text{ and } B) = P(A) \cdot P(B)$, when A and B are independent events.

In order to show the logic for taking the product of the two individual probabilities described in the above situation, let's look again at what it means for two events to be independent. The following illustration is shown graphically by a Venn diagram for the purpose of demonstrating this relationship. Events A and B are defined on a sample space of six sample points. Three of the six belong to set A, the solution set for event A. Two points belong to set B, and so on, as shown on the Venn diagram (Figure 4-4). (Each point has been numbered for identification.)

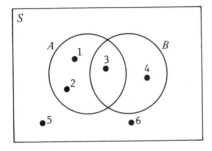

Figure 4-4

Note: Event $A = \{1, 2, 3\}$
Event $B = \{3, 4\}$
Event A and $B = \{3\}$

$P(A) = 3/6 = 1/2$ and $P(B) = 2/6 = 1/3$. I claimed that these events were independent. Are they? $P(B|A) = P(B, \text{ knowing that } A \text{ has happened}) = 1/3$, as shown in the unshaded portion of the Venn diagram in Figure 4-5. Since $P(B) = P(B|A)$, then the two events are independent.

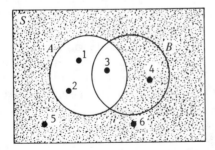

Figure 4-5

The probability of (A and B) is represented by the proportion of the sample space that lies in the intersection of sets A and B. When we multiply the probability of event A by the probability of event B, (1/2)(1/3), we are in effect finding what proportion of the whole sample space 1/3 of 1/2 represents. That is, we know that 1/2 of the sample space belongs to set A. We also know that 1/3 of set A belongs to set B. Thus, we are looking for 1/3 of the 1/2, or 1/6. You will notice that the set $A \cap B$ (point 3) represents exactly 1/6 of the sample space.

Let's consider another situation: two dependent events, C and D, as shown in Figure 4-6. If we multiply these two values, we find that $P(C) \cdot P(D) = (3/8)(3/8) = 9/64$. However, by inspection we see that P(C and D) is 1/8 and $1/8 \neq 9/64$. Since these two values are not the same we can be sure that the 9/64 is incorrect. (The 1/8 was found by definition.) The multiplication rule does not hold for this example, because the two events, C and D, are not independent. The conditional probability of event C, given that D has occurred, P(C|D), is 1/3, as shown on the unshaded portion of Figure 4-7.

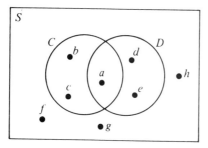

Figure 4-6

$P(C) = 3/8$ and $P(D) = 3/8$

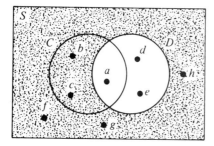

Figure 4-7

Since the two probabilities are not the same in value [$P(C) = 3/8$ and $P(C|D) = 1/3$], we must conclude that the two events, C and D, are dependent events. When this occurs we have no right to apply the multiplication rule, as stated above.

When the events involved are dependent, as with C and D above, the second form of the multiplication rule can be applied.

The Multiplication Rule (Case 2) If events A and B are dependent events defined on a sample space, then the probability that both occur simultaneously is found by multiplying the probability of one event by the conditional probability of the other.

$$P(A \text{ and } B) = P(A) \cdot P(B|A)$$
or
$$P(A \text{ and } B) = P(B) \cdot P(A|B).$$

Apply either one of these statements using the conditional probability that you have previously obtained or can obtain. Returning to our above illustration involving events C and D, we find that we have previously found $P(C|D)$ to be 1/3. Therefore we will find the probability of $(C$ and $D)$ by multiplying $P(C|D)$ and $P(D)$.

$$P(C \text{ and } D) = P(D) \cdot P(C|D)$$
$$= 3/8 \cdot 1/3 = \underline{1/8}$$

This value agrees with the value found previously by definition.

The multiplication rule is convenient to use for finding the probability of an "and" event provided you know or can determine which case confronts you: case 1, independent events, or case 2, dependent events. On occasion, however, there will be situations where it is difficult to determine which case to use. Recall that it is always possible to use the basic formula for probability, $P(\text{any event}) = [n(\text{any event})]/n(S)$, provided you have an equally likely sample space.

EXERCISES 4-4

1. Events R and S are independent and their probabilities are $P(R) = 1/4$ and $P(S) = 1/2$.
   a. Find the probability of $(R$ and $S)$.
   b. Draw a Venn diagram that shows this relationship, and check, by using probabilities, to see if your diagram shows independent events.

2. Given that $P(A) = 1/3$, $P(B) = 1/4$, and $P(A \text{ and } B) = 1/5$, determine whether or not A and B are independent events. Demonstrate the reasons for your answer with the aid of a Venn diagram and probabilities.

3. Using events F and K as defined in Exercise 4 of Lesson 4-3, find the probability that the card drawn satisfies both events F and K, that is, $P(F \text{ and } K)$, in two ways.

LESSON 4-5: THE ADDITION RULE

The addition rule, like the multiplication rule, is a formula for calculating the probability of a compound event. The addition rule is used when the connective

ADDITION RULE   Lesson 4-5

is "or." The following example, with the aid of a Venn diagram, will show the use of the basic addition rule, which is:

$$P(A \text{ or } B) = P(A) + P(B) - P(A \text{ and } B)$$
[formula (4-4a), p. 157 in textbook]

The solution set for an "or" event is the union of the solution sets that correspond to the individual events. In Figure 4-8 we see two events, R and T. The probability measure assigned to set R is the sum of the probability measures of the two regions a and b (shaded portion). The probability measure of set T is the sum of the probability measure of regions a and c, as shown in Figure 4-9. The probability measure associated with the union of sets R and T would be the sum of the measures associated with the three regions a, b, and c.

We would like a rule that would allow us to combine the probability $P(R)$ [the sum of the probability measures of regions a and b] with the probability $P(T)$ [the sum of the probability measures of regions a and c]. If these two probabilities are simply added, the probability measure of region a will be included twice. You will recall that we have seen this kind of situation before (in Lessons I-7 and 4-2), and the element belonging to this region cannot be accounted for twice; thus a "correction factor" must be used. The probability measure of region a is then subtracted from the total.

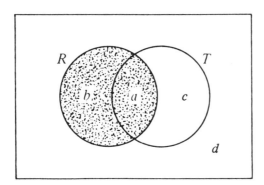

Figure 4-8

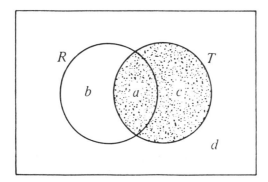

Figure 4-9

Illustration Consider a probability experiment where two events, A and B, are defined. Let $P(A) = 1/3$, $P(B) = 1/4$, and $P(A \text{ and } B) = 1/5$. Find the probability that event (A or B) will occur when the experiment is tried.

Solution
$$P(A \text{ or } B) = P(A) + P(B) - P(A \text{ and } B)$$
$$= (1/3) + (1/4) - (1/5) = (20/60) + (15/60) - (12/60)$$
$$= \underline{23/60}$$

Consider the Venn diagram shown in Figure 4-10. The Union, $A \cup B$, is composed three separate cells. The sum of the three probabilities is 23/60, the same value as found when using the addition-rule formula.

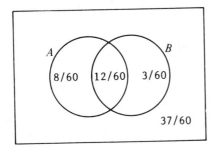

Figure 4-10

In the above illustration, P(A and B) was given information. If P(A and B) is not given by the problem, then it must be found separately. The probability of the "and" compound event was discussed previously in two forms: (1) in Lesson 4-2 as a sample space event, where the definition of probability applied directly, and (2) in Lesson 4-4, where we investigated the use of multiplication to calculate the probability of a statement formed by "and." If the situation includes independent events, then the multiplication rule can be used for the last term of the equation for the addition rule.

Illustration Suppose we have two events, F and Q, defined on a sample space. $P(F) = 1/2$ and $P(Q) = 2/3$, and they are independent events. Find the probability of $P(F \text{ or } Q)$.

Solution The addition rule: $P(F \text{ or } Q) = P(F) + P(Q) - P(F \text{ and } Q)$. $P(F \text{ or } Q) = (1/2) + (2/3) - P(F \text{ and } Q)$. But since F and Q are independent, $P(F \text{ and } Q) = (1/2)(2/3) = 1/3$. Therefore: $P(F \text{ or } Q) = (1/2) + (2/3) - (1/3) = \underline{5/6}$.

If a problem does not specify whether or not the events are independent, you will be expected to determine P(A and B) in the best way possible. Determine whether or not they are independent, or work with the sample space and find the appropriate relative frequencies to test.

The use of the addition rule is greatly simplified when the individual events are mutually exclusive: the probability measure of the union of mutually ex-

SELF-CORRECTING EXERCISES

clusive events is the sum of the probabilities of the individual events. This case is discussed and illustrated in the textbook on pages 156-159.

EXERCISES 4-5

1. Find $P(A$ or $B)$ given that A and B are independent events defined on the same sample space, and that $P(A) = 0.3$ and $P(B) = 0.4$.

2. Find $P(C$ or $D)$, given that events C and D are mutually exclusive events, and that $P(C) = 0.75$ and $P(D) = 0.125$.

3. Using events F (face card) and K (king) as defined in Exercise 4 of Lesson 4-3, find each of the following.
   a. $P(F$ or $K)$ b. $P(F$ or $\bar{K})$ c. $P(\bar{F}$ or $K)$

SELF-CORRECTING EXERCISES FOR CHAPTER 4

1. From a list of 5 foods (A, B, C, D, E) a person is asked to select the 2 he or she most prefers. Write a sample space showing the possible responses to this experiment.

2. The numbers 0, 1, 2, 3, 4, 5, 6, 7, 8, 9 are put on identical cards, placed in a box, and thoroughly mixed. One card is drawn at random. What is the probability that the card drawn:
   a. has the numeral 3 on it?
   b. has an even numeral on it?
   c. has a number less than 8 on it?
   d. has a number less than 8 that is also odd on it?
   e. has a number that is either odd or less than 5 on it?

3. A single card is drawn from a well-shuffled deck of 52 playing cards. What is the probability that it is a jack?

4. A single letter is picked randomly from the English alphabet.
   a. What is the probability that it is a vowel?
   b. What is the probability that it is one of the last six letters of the alphabet?

5. If a month of the year is to be selected at random, what is the probability that it has exactly 30 days?

6. A pair of dice is rolled. What is the probability that:
   a. a total of 11 shows?
   b. a total of 7 or a total of 11 shows?
   c. a total of 6 does not show?
   d. a total of more than 8 shows?
   e. a total of less than 18 shows?

82 PROBABILITY Chapter 4

7. The histogram below represents data from a sample of 50 families; the variable is the number of children in each family. If one family is selected at random from this sample, find the probability that:
   a. The family has 2 children.
   b. The family has 4 or more children.
   c. (true or false) The individual bars of this histogram represent a set of mutually exclusive events.

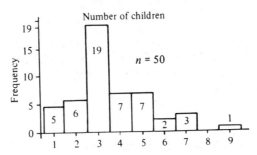

8. A box contains 3 poker chips, 1 red, 1 white, and 1 blue. Two chips are drawn, one at a time, with the first being replaced before the second is drawn.
   a. List the sample space and assign a probability to each sample point.
   b. What is the probability that both are white?
   c. What is the probability that both are blue?
   d. What is the probability that both are the same color?
   e. What is the probability that neither is white?
   f. What is the probability that at least one is blue?

9. In a room of 25 people, 5 are color blind. If 2 people are selected at random, what is the probability that:
   a. both are color blind?
   b. neither is color blind?

10. A box contains 20 identical-looking flashlight batteries; 15 are good and 5 are bad.
    a. If one battery is selected at random, what is the probability that it is good?
    b. If 3 are selected at random (without replacement), what is the probability that all 3 are good?

11. A box contains 6 poker chips, 1 red, 2 white, and 3 blue. Two poker chips are drawn from the box without replacement. (Think of drawing them one at a time.)
    a. Find the sample space and assign a probability to each sample point.
    b. Find the probability that both are red.
    c. Find the probability that both are white.
    d. Find the probability that both are blue.
    e. Find the probability that the first one drawn is red.

SELF-CORRECTING EXERCISES

  f. Find the probability that the second one drawn is red.
  g. Find the probability that neither one is red.

12. One letter is to be selected at random from each of two words: TOOT and BOOT. (Think of each letter of each word as being on a separate card and each word in a separate box. One letter is then drawn from each box.)
    a. List an appropriate sample space for this experiment and assign a probability to each sample point.
    b. What is the probability that the same letter is drawn from both words?

13. An experiment consists of drawing two bills from a box containing 5 one-dollar bills and 2 ten-dollar bills. (Think of drawing one and then a second without replacing the first.) What is the probability that:
    a. both are ten-dollar bills?
    b. both are one-dollar bills?
    c. one of each is drawn?

14. An experiment consists of flipping a balanced coin until a head occurs for a maximum of three times.
    a. List the sample space for this experiment and assign the appropriate probabilities to each sample point.

    Suppose event H is defined as "a head occurred" and event T is defined as "three tosses were required."
    b. Are H and T mutually exclusive events? Explain.
    c. Are H and T independent events? Explain.

15. A two-stage experiment consists of selecting one bag from two identical bags and then randomly selecting one piece of fruit from that bag. Bag I contains 6 oranges and 4 peaches, while bag II contains 3 oranges and 6 peaches. What is the probability that:
    a. an orange is selected?
    b. a peach is selected?
    c. an orange is selected from bag II?
    d. an orange is selected, knowing that bag I was chosen?
    e. the fruit came from bag II, knowing that an orange resulted?

16. A large shipment of grapefruit contains:

    10% white seedless
    20% pink seedless
    30% pink with seeds
    40% white with seeds

    If one grapefruit is picked at random from this shipment, what is the probability that:
    a. it has seeds?
    b. it is seedless?

c. it is white?
d. it is pink?
e. it is pink and seedless?
f. it is pink or has seeds?
g. it is pink, knowing that it has seeds?
h. it has seeds, knowing that it is pink?

17. An experiment consists of two persons ranking a new record album. They each rank the album on a scale from 1 to 4. The sample space is given below and three events, *S*, *F*, and *T* are defined on this sample space. (To answer the probability questions below, assume that each rank number is as likely to be used as each other one.)

S: both ranked album the same
F: album received at least one ranking of four
T: the total of the two rankings is four or less

		1st person			
		1	2	3	4
2nd person	1	1,1	2,1	3,1	4,1
	2	1,2	2,2	3,2	4,2
	3	1,3	2,3	3,3	4,3
	4	1,4	2,4	3,4	4,4

Find the following probabilities.
a. *P*(*S*)   b. *P*(*F*)   c. *P*(*T*)
d. *P*(*S* and *F*)   e. *P*(*F* and *T*)   f. *P*(*S* and *T*)
g. *P*(*S* or *F*)   h. *P*(*F* or *T*)   i. *P*(*S* or *T*)
j. Which pairs of events are mutually exclusive: (*S*,*F*), (*S*, *T*), or (*F*,*T*)?

18. If $P(A) = 1/2$ and $P(B) = 2/5$, and if *A* and *B* are independent events, then find each of the following.
a. *P*(*A* and *B*)   b. *P*(not *B*)   c. *P*(*A* or *B*)

19. If events *A* and *B* are mutually exclusive events, and if $P(A) = 0.5$ and $P(B) = 0.3$, find each of the following.
a. *P*(*A* or *B*)   b. *P*(*A* and *B*)

20. If events *C* and *D* are independent events, and if $P(C) = 0.7$ and $P(D) = 0.2$, find each of the following.
a. *P*(*C* or *D*)   b. *P*(*C* and *D*)

21. Of the patients examined at a clinic, 0.3 had lung-related difficulties, 0.4 had blood deficiency, and 0.1 had both.
a. What is the probability that a patient selected at random had at least one of these problems?
b. Are the two events, "lung-related difficulties" and "blood deficiency," independent events? Justify your answer.

SELF-CORRECTING EXERCISES

22. Adam and Bob both opened new businesses at the same time. The probability that each is a great success during the first year is 0.4 and 0.2, respectively. If the success of each business is independent of the other, find the probability that:
    a. both will be a great success during the first year.
    b. neither will be a great success during the first year.
    c. at least one will be a great success during the first year.

23. An aluminum company is bidding on two parcels of land, A and B. Its executives feel that the probabilities of winning the bids on A and B are 0.5 and 0.6, respectively. It is also believed that the probability of A having aluminum ore is 0.2 and of B having aluminum ore is 0.1. What is the probability that the company will obtain more aluminum ore as a result of these bids. Assume independence.

24. Company A is planning to market a new product. If its competitor does not market a similar product, then the probability that the product will do well is 0.9. If the competitor develops and markets a similar product, the probability that company A's product will do well is 0.5. It is believed that the probability of the competitor marketing a similar product is 0.6. What is the probability that Company A's product will do well?

PROBABILITY DISTRIBUTIONS (DISCRETE VARIABLES)

5

LESSON 5-1: THE MEAN AND STANDARD DEVIATION OF A DISCRETE PROBABILITY DISTRIBUTION

A discrete probability distribution can be thought of as a relative frequency distribution for an entire population. Since the probability distribution is a representation of the entire population, its mean and standard deviation are parameters.

Recall that the symbol used to represent the measures of a population is different from the one used for a sample (see page 8 in the textbook). A number that describes some property of a sample is called a "sample statistic," whereas a number that describes a characteristic of the population is called a "population parameter." The sample has a mean, $\bar{x}$, as does the population, only the population mean is identified by μ (the lowercase Greek letter "mu"). The sample standard deviation is identified by s, while the standard deviation of the population is identified by σ (the lowercase Greek letter "sigma").

As indicated on page 176 in the textbook, the mean of the population can be found in the same manner as the mean of a sample: Sum up all the values of x and divide by n. Below you will find an illustration where the mean of the population is found in both ways. You will see that the two methods are comparable and that the formula for μ is a reasonable alternative for the calculation of this mean.

Consider the theoretical probability distribution described by $P(x) = x/10$, for $x = 1, 2, 3, 4$ (see Table 5-1).

Table 5-1

x	1	2	3	4
$P(x)$	1/10	2/10	3/10	4/10

If a sample of size 10 were to be taken from this population we would expect the following frequency distribution to result (Table 5-2).

MEAN AND STANDARD DEVIATION Lesson 5-1

Table 5-2

x	1	2	3	4
f	1	2	3	4

The mean of this distribution would be calculated by using formula (2-2), found in the textbook on page 50. (See Table 5-3 below.)

Table 5-3

x	f	xf
1	1	1
2	2	4
3	3	9
4	4	16
	10	30

$$\bar{x} = \frac{\Sigma xf}{n}, \text{ where } n = \Sigma f$$

$$\bar{x} = \frac{30}{10} = 3.0$$

If the relative frequency distribution (the probability distribution) were used [f/n or $P(x)$ in place of f], we would see the same end result. The difference is that each of the addends would have a common denominator. We would be summing fractions instead of adding just the numerators and then dividing by the n, 10 in this case (Table 5-4).

Table 5-4

x	P(x) or f/n	x · P(x) or x · (f/n)
1	1/10	1/10
2	2/10	4/10
3	3/10	9/10
4	4/10	16/10
	10/10 = 1.0	30/10 = 3.0

Notice the similarity in the two tables. Thus for this example, we see that $\Sigma x \cdot P(x)$ has exactly the same value as $\Sigma xf/n$. However, since a probability distribution is used only to represent a population, the mean of the population, μ, will be found by:

$$\mu = \Sigma[x \cdot P(x)] \qquad \text{[formula (5-2), p. 197 in textbook]}$$

Note: The above example was used solely to demonstrate that the formula for calculating the mean of a discrete probability population was consistent with the ideas and formulas presented in Chapter 2. The example below used for the calculation of the standard deviation serves the same purpose.

Let's again consider the expected sample of size 10 from our discrete probability population, where $P(x) = x/10$ for $x = 1, 2, 3,$ and 4. Now recall that the definition for the standard deviation of a *sample* was:

$$s = \sqrt{\frac{\Sigma(x - \bar{x})^2}{n - 1}}$$ [formula (2-8), p. 61 in textbook]

Recall also that in Lesson 2-2 we commented specifically on the denominator, $n - 1$. The number $n - 1$ will be the denominator used when the set of data available is a sample of some population. However, if we had the entire population, the denominator would be the number of pieces of data, n. Since we are using a collection of ten numbers to represent a given population, we should make this adjustment. The formula for the standard deviation of a population would then be:

$$\sigma = \sqrt{\frac{\Sigma(x - \mu)^2}{n}}$$

Recall that the mean and standard deviation for a population are represented by μ and σ, respectively. This formula could be expressed in the form of the shortcut formula (3-2) as:

$$\sigma = \sqrt{\frac{n(\Sigma x^2 f) - (\Sigma xf)^2}{n(n)}}$$

Let's apply this formula to our "expected sample." (See Table 5-5.)

Table 5-5

x	f	xf	$x^2 f$
1	1	1	1
2	2	4	8
3	3	9	27
4	4	16	64
	10	30	100

$$\sigma = \sqrt{\frac{(10)(100) - (30)(30)}{10(10)}} = \sqrt{\frac{1000 - 900}{100}} = \sqrt{\frac{100}{100}} = 1.0$$

A simple rewriting of this formula will produce formula (5-4a) as shown in the textbook on page 197.

$$\sigma = \sqrt{\frac{n \cdot \Sigma x^2 f}{n \cdot n} - \frac{\Sigma xf \cdot \Sigma xf}{n \cdot n}} = \sqrt{\left[\frac{\Sigma x^2 f}{n}\right] - \left[\frac{\Sigma xf}{n}\right]\left[\frac{\Sigma xf}{n}\right]}$$

Recall that earlier in this lesson we were able to adjust the denominator n,

BINOMIAL PROBABILITIES Lesson 5-2

since it is a constant, so as to convert f/n into a probability. We can do the same thing here:

$$\sigma = \sqrt{\Sigma x^2 \cdot P(x) - [\Sigma x \cdot P(x)]^2}$$

Let's take a second look at our example now (Table 5-6).

Table 5-6

x	$P(x)$	$x \cdot P(x)$	$x^2 \cdot P(x)$
1	1/10	1/10	1/10
2	2/10	4/10	8/10
3	3/10	9/10	27/10
4	4/10	16/10	64/10
	$\frac{10}{10} = 1.0$	$\frac{30}{10} = 3.0$	$\frac{100}{10} = 10$

$\sigma = \sqrt{10 - (3)^2}$

$\sigma = \sqrt{1} = \underline{1.0}$

Notice again that the only real difference between the two calculations is the location of the denominator.

Be aware of the distinction between "any sample" and the "expected sample" that was used in the above illustration. The "expected sample" is really the smallest collection of numbers (1, 2, 2, 3, 3, 3, 4, 4, 4, 4) that could represent the discrete population used for the illustration.

EXERCISES 5-1

1. a. Identify each of these four symbols: $\bar{x}$, s, μ, and σ.
   b. Explain the difference between $\bar{x}$ and μ.
   c. Explain the difference between s and σ.

2. a. Calculate the mean (μ) and the standard deviation (σ) for the discrete population described by $P(x) = 5 - x/10$ for $x = 1, 2, 3, 4$.
   b. Compare this population to the one in our example.

3. Find the mean and standard deviation of the probability distribution below.

x	1	2	4	5	9	all other x's
$P(x)$	1/9	2/9	3/9	2/9	1/9	0

LESSON 5-2: BINOMIAL PROBABILITIES

The binomial probability formulas and tables can be used only when the experiment in question is binomial. We can easily see that the tossing of a coin is binomial since it has two possible outcomes, heads or tails. There are, however, numerous everyday situations that also qualify. To name a few: (1) I did or did not have an automobile accident today; (2) the flash bulb worked or did not work; (3) the jet flight arrived or did not arrive on time; (4) the laboratory mouse has or has not learned his required reaction to a set stimulus; (5) this newly manufactured item, fresh off the production line, does or does not perform its designed function. The list is endless.

Each of these probability experiments demonstrates the basic properties of a binomial experiment, just as coin tossing does. First, each trial has outcomes that can be classified into one of two categories, success or failure. (Usually "success" is the specific outcome that we are looking for, the outcome whose occurrences we are counting. Failure is anything else.) Second, each experiment requires several repeated trials. These repeated trials must be independent. (Independent, as discussed in Lesson 4-3, means that the probability of success, P, remains at a constant value throughout the experiment.)

Let's investigate the binomial probability function, $P(x) = \binom{n}{x} p^x q^{n-x}$ for $x = 0, 1, 2, 3, \ldots, n$ [formula (5-6) on page 206 of the textbook], by looking at a specific illustration. A second-grade class, while studying plant growth, plants some vegetable seeds. Each child in the class plants six seeds. The seed company says that 90 percent of these seeds will germinate. How does this class experiment relate to a binomial experiment?

Each child's set of six seeds constitutes a binomial experiment. Each seed within one child's set will be an individual trial — it will grow (success) or it will not grow (failure). Each individual seed is planted separately and cared for as directed by the packer; thus we will assume that the individual trials are independent. The random variable x will be the number of seeds that germinate for each child. All probability questions that might be asked about this situation must be broken down and expressed in terms of the individual values of x. For example, what is the probability that at least half of a given child's seeds germinate? Before we answer that and other questions, let's look at a single situation and inspect our probability function.

This experiment is a binomial experiment since each seed will grow or not grow and each of the six seeds is a separate trial. (Each child's plantings represent a different experiment.) The probability of germination for each seed is assumed to be 0.90 ($p = 0.90$), while x is the number of seeds, within each child's experiment, that do germinate. Notice that x could be any one of the numerical values 0, 1, 2, 3, 4, 5, or 6. Let's consider the case where $x = 4$; that is, exactly four seeds germinate for a particular child. The probability that exactly four germinate is related to (0.9)(0.9)(0.9)(0.9)(0.1)(0.1), which is the product of the probabilities that germination occurs with four seeds and does not occur with the other two seeds. However, $P(x = 4)$ does not equal $(0.9)^4 (0.1)^2$. More specifically, this is the probability that the first four are the four that germinate, and the last two are the two that do not germinate. Let S and F represent "do germinate" and "do not germinate," respectively. (Recall that x is only the

BINOMIAL PROBABILITIES   Lesson 5-2

number that do germinate.) The above product represents the probability of SSSSFF. But $x = 4$ for several other possible arrangements. For example, $x = 4$ for all of the following cases:

SSSSFF				
SSSFSF	SSSFFS			
SSFSSF	SSFFSS	SSFSFS		
SFSSSF	SFFSSS	SFSFSS	SFSSFS	
FSSSSF	FFSSSS	FSFSSS	FSSFSS	FSSSFS

There are fifteen specific ways that a set of six seeds could have exactly four seeds germinate. The probability for each one of these specific arrangements is $[(0.9)^4(0.1)^2]$. The probability for $x = 4$ must be the total of all fifteen, or $(15)[(0.9)^4(0.1)^2] \approx 0.098$. (The symbol $\approx$ means "nearly equal to.") This says that the probability that a given child's experiment has exactly four seeds germinate is approximately 0.098. (This probability may be calculated by using the formula or it may be obtained from Table 4 on page 570 in the textbook.)

The probability that at least half the seeds of a child germinate may be found by adding the appropriate probabilities.

P(at least half germinate) $= P(x = 3, 4, 5,$ or $6)$ (half or more)
$= P(x = 3) + P(4) + P(5) + P(6)$

We will add these separate probabilities since the values of $x = 3, 4, 5,$ and 6 are mutually exclusive (addition rule).

P(at least half germinate) $= 0.015 + 0.098 + 0.354 + 0.531 = \underline{0.998}$

(These values were obtained from Table 4, page 570 in the textbook, by using $n = 6, p = 0.90$, and the required values for x.)

Notes: 1. We can multiply the probabilities of independent events (or trials) to obtain the probability of a compound event formed by the connective "and."

2. We can add the probabilities of mutually exclusive events to obtain the probability of a compound event formed by the use of the connective "or."

3. The number 15, the number of all the different cases that contain exactly four successes in our illustration, can be obtained by the use of the binomial coefficient. In our example, $n = 6$ and $x = 4$, and the value of $\binom{n}{x}$ is 15. By definition, this symbol represents the number of different arrangements that are possible and that still have exactly x successes in the allotted n trials. See Lesson I-9, for additional information about $\binom{n}{x}$.

EXERCISES 5-2

1. An inspector at the end of a production line opens cases for inspection. Each case contains twelve items. The manufacturer claims that approximately 1 percent of his items are defective.
   a. Describe the specifics of this situation that make it a binomial experiment.
   b. Find the probability that there are no defectives in a given case.
   c. Find the probability that no more than one defective is found.

2. John has not studied for the quiz today, so he decides to guess at the answers without reading either the questions or the answers. The quiz consists of ten multiple-choice questions with five possible answers, only one of which is correct.
   a. What is the probability that John will guess exactly five correct answers?
   b. What is the probability that he gets less than half of the answers correct?
   c. Find the probability that he gets exactly two correct answers.
   d. Make a list showing all the possible ways that exactly two answers are correct (identify the two correct answers for each case).
   e. What is the probability that any one of the specific situations [answers to part d] occurs?
   f. How can answers to parts d and e be used to obtain the answer to part c?

SELF-CORRECTING EXERCISES FOR CHAPTER 5

1. From the sample space of the record album rankings in Exercise 17 (self-correcting section of Chapter 4), construct a probability distribution for the random variable x, the sum of the two rankings.

x	$P(x)$
2	1/16
.	
.	
.	

2. Calculate the mean and standard deviation for the probability distribution of x.

x	$P(x)$
1	.1
3	.2
5	.4
7	.2
9	.1

SELF-CORRECTING EXERCISES

3. Find the mean and the standard deviation for the following probability distribution.

x	P(x)
1	.1
2	.2
3	.3
4	.2
5	.2

4. Given the probability distribution.

x	P(x)
3	0.1
4	0.2
5	0.3
6	0.4

    a. Find μ.
    b. Find σ.

5. Calculate the probability that all coins will show heads if:
    a. two coins are tossed.
    b. three coins are tossed.
    c. four coins are tossed.
    d. ten coins are tossed.

6. A large shipment of walkie-talkies is being inspected by checking five of them picked at random. If all the inspected ones work properly, the shipment will be accepted; otherwise, the shipment will be rejected.
    a. What is the probability of accepting the shipment if it is actually 10 percent defective?
    b. What is the probability of accepting the shipment if it is actually 20 percent defective?

7. Three identical coins are flipped simultaneously. Find the probability that:
    a. all coins show heads.
    b. exactly two coins show heads.
    c. exactly one coin shows a head.
    d. exactly none of the coins show heads.
    e. at least one head is observed.
    f. all coins show the same result.

8. Ten percent of the male students at our college are veterans. Two male students are selected at random from the college. What is the probability that:

a. both are veterans?
b. neither is a veteran?
c. exactly one of them is a veteran?

9. An archer has a history of hitting the bull's-eye 80 percent of the time. Find the probability that he hits the bull's-eye:
   a. all eight times in his next eight tries.
   b. at least seven times in his next eight tries.
   c. at least six times in his next eight tries.

10. If the proportion of male children is 1/2, find the probability that in a family of six children (assume independence):
    a. all six are boys.
    b. the youngest is a girl and the other five are boys.
    c. one is a girl and the other five are boys.
    d. two are girls and the other four are boys.
    e. three are girls and three are boys.
    f. Explain why the situation described above satisfies the conditions of a binomial experiment.

11. A coin is loaded such that a tail is three times as likely to occur as a head. The coin is flipped twice.
    a. Find the probability that two heads occur.
    b. Find the number of heads expected to be observed.

12. Given the probability that a patient's full recovery from a certain disease is 0.6, find each of the following.
    a. The probability that exactly three of a set of four patients will recover. (Assume independence.)
    b. Calculate the mean number of patients that will recover for each set of four patients under the above conditions.
    c. Calculate the standard deviation of the number of patients that recover for each group of four patients.

13. A coin is loaded such that the tail is twice as likely to occur as the head. An experiment consists of tossing this coin three times and observing the number of heads that occur.
    a. Construct a table showing the probability distribution for this experiment.
    b. Draw a histogram depicting this distribution.
    c. Using the distribution found in part a, calculate the mean and standard deviation for this experiment using formulas (5-2) and (5-4) from the textbook.
    d. The experiment above is binomial. Justify this by describing each property as it applies to this experiment.
    e. Calculate the mean and the standard deviation using the binomial formulas (5-7) and (5-8) from the textbook.

SELF-CORRECTING EXERCISES

14. A die is loaded such that the 6 is twice as likely to occur as any of the other numbers. 1, 2, 3, 4, or 5 are equally likely. The experiment is to roll this die once and observe the number of spots on the top surface.
    a. Construct a table showing the probability distribution for this experiment.
    b. Draw a histogram depicting this distribution.
    c. Using the distribution found in part a, calculate the mean and standard deviation for this experiment using formulas (5-2) and (5-4) from the textbook.
    d. This experiment is not binomial. Explain why not.

THE NORMAL PROBABILITY DISTRIBUTION

6

LESSON 6-1: THE STANDARD SCORE

The standard score is the name given to the numerical value assigned to the random variable when it is "standardized." This "standardizing" process is merely a matter of converting the unit of measure for any given distribution of data to a scale where (1) the origin (zero) is at the mean value of the distribution and (2) the length of one unit is always the size of one standard deviation of that same distribution. In Figure 6-1 you will see a distribution of x which has been converted to its corresponding "standard" units. Thus the standard score for any given value of a random variable is the number of standard deviations the value is above or below the mean value. (This could be considered a form of coding.)

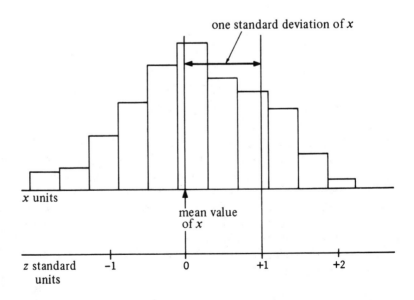

Figure 6-1

STANDARD SCORE Lesson 6-1

A standard score, z, of $+1.0$ means that the corresponding value of the random variable is exactly one standard deviation larger than the mean value ($z = +1.0$ when $x = \mu +$ one standard deviation). A standard score of -1.2 corresponds to the value which is 1.2 standard deviations smaller than the mean ($z = -1.2$ when $x = \mu - 1.2\sigma$).

The formula for calculating the z score that corresponds to any given value of a random variable x is:

$$z = \frac{x - \mu}{\sigma}$$ [formula (6-3), p. 224 in textbook]

where μ is the mean value of the x distribution and σ is the standard deviation of this same distribution of x-values.

Illustration Find the z score that corresponds to $x = 57.2$ in a distribution where the mean is 52.0 and the standard deviation is 2.5.

Solution $x = 57.2, \mu = 52.0, \sigma = 2.5$.

$$z = \frac{x - \mu}{\sigma} = \frac{57.2 - 52.0}{2.5} = \frac{5.2}{2.5} = \underline{2.08}$$

Illustration Find the value of x that corresponds to $z = -2.35$ in a distribution where the mean is 62.5 and the standard deviation is 8.4.

Solution $\mu = 62.5, \sigma = 8.4, z = -2.35$.

$$z = \frac{x - \mu}{\sigma}$$

$$-2.35 = \frac{x - 62.5}{8.4}$$

$$x - 62.5 = (-2.35)(8.4)$$

$$x = 62.5 - 19.74 = \underline{42.76}$$

The main use that we will have for the standard score will be in connection with the normal distribution. A refinement of the empirical rule gives us information about the normal distribution in terms of standard units. If some variable does in fact have a normal distribution, then its distribution has exactly the same properties as any other normally distributed random variable. The only difference between the two distributions is the location of the mean and the size of the standard deviation. This application of the standard score to the normal distribution is discussed and illustrated in the textbook, pages 224-228, and in the next lesson, 6-2.

EXERCISES 6-1

1. Find the standard score that corresponds to each of the following values of x, if they belong to a distribution where the mean is 27.8 and the standard deviation is 4.2.
   a. $x = 27.8$ b. $x = 32.0$ c. $x = 35.2$
   d. $x = 25.4$ e. $x = 12.1$ f. $x = 29.4$

2. True or false.
   a. The standard score will always take on a value between -3.0 and +3.0.
   b. The standard score can be used only with a normal distribution.

3. Find the value of x that corresponds to the value of z, the standard score, given below. Assume that all values refer to a distribution where the mean is 125.9 and the standard deviation is 8.7.
   a. $z = 1.2$ b. $z = 1.5$ c. $z = 0.5$
   d. $z = -2.4$ e. $z = 4.25$ f. $z = -3.75$

LESSON 6-2: PROBABILITIES ASSOCIATED WITH THE NORMAL DISTRIBUTION

The probability of an event associated with a normal distribution will be found by using the standard normal distribution. The probabilities related to the standard normal distribution are given in Table 5 on page 573 in the textbook. The standard normal distribution is a normal distribution of the standard score z. (The mean is zero and the standard deviation is one unit.) All other normal distributions then become transformations of this standard distribution.

Table 5 is so constructed that we can read the probability measure of the area between the mean ($z = 0$) and some standard score above the mean, correct to the nearest hundredth. That is, the number in the table represents the probability measure from $z = 0$ up to the specific z score that identifies the measure obtained. The units and tenths digits are located along the left side of the table; the hundredths digit is located along the top. A z score of 1.42 identifies a probability of 0.4222. Figure 6-2 illustrates this example. The textbook contains a complete description of how the various probabilities are obtained for the standard score and also how the probabilities associated with any normally distributed random variable are found.

z	...	.02
⋮		
1.4		.4222

Figure 6-2

PROBABILITIES FOR THE NORMAL DISTRIBUTION Lesson 6-2

In this lesson, we should like to point out some miscellaneous details from various chapters and show you their interrelationship.

The mean of the normal distribution is the same as the median since the distribution is symmetrical. Therefore, exactly one-half of the distribution lies to the right (or to the left) of the mean value. We will take advantage of this property frequently during the use of the normal distribution.

The application of the rules for calculating probabilities of a normally distributed random variable frequently requires the use of addition or subtraction or both. (Subtraction is really a form of addition.) The use of addition is perfectly justifiable when the probabilities that are being added (or subtracted) are the probabilities of mutually exclusive events. Recall that in Chapter 4 in the textbook we were introduced to an "addition rule," one form of which said that $P(A \text{ or } B) = P(A) + P(B)$, provided A and B were mutually exclusive events. This rule will be applied often in the use of the normal distribution table. For example, the probability that z is greater than 1.5 requires that we subtract 0.4332 from 0.5000 (0.5000 - 0.4332 = 0.0668) to obtain the answer, $P(z > 1.5) = 0.0668$. Figure 6-3 suggests why we subtract.

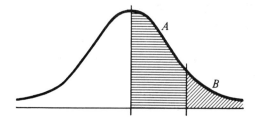

Figure 6-3

One might look at this from a different point of view. $P(z > 0) = 0.500$ is the measure of all the area under the curve to the right of the mean. This area is split into two sections: one part is between $z = 0$ (the mean) and $z = 1.5$, and the other is above (to the right of) $z = 1.5$. The measures of these two sections are represented by the two shaded regions in the diagram and are labeled regions A and B, respectively. Clearly the sum of the two regions is the total of the area to the right of the mean; thus, $P(z > 0) = P(0 < z < 1.5) + P(z > 1.5)$. The probability $P(0 < z < 1.5)$ is the 0.4332 read from Table 5. $P(z > 0)$ is known to be 0.5000, as discussed above. Therefore, a simple rewriting of the above equation gives us the solution for the probability that z is greater than 1.5: $P(z > 1.5) = 0.5000 - 0.4332$, or $P(z > 1.5) = 0.0668$, as stated above. (Since z cannot be both larger and smaller than 1.5 at the same time, these two events are mutually exclusive and the addition rule holds as shown.) The other situations where we add or subtract can be justified by the same line of reasoning.

The property of symmetry is also used a great deal in calculating the probabilities of regions on the left half of the distribution. Symmetry means that one side of a distribution is a mirror-like reflection of the other side, divided at the mean value. For example, it is symmetry that enables us to find the probability that z is less than −1.5. Symmetry shows us that the value of $P(z < 1.5)$ is exactly the same as the value of $P(z > +1.5)$.

If Figure 6-4 were folded at the mean, we would see that these two regions fit exactly on top of one another. If they are symmetrical, then their measures are the same; and $P(z < -1.5)$ is 0.0668, as was $P(z > +1.5)$.

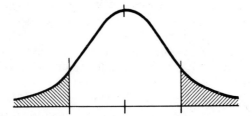

Figure 6-4

EXERCISES 6-2

1. Find the probability in Table 5 that corresponds to the z score listed below.
   a. 0.74   b. 1.06   c. 2.15
   d. 1.88   e. 3.24   f. 4.00

2. Find each of the following probabilities using the property of symmetry and operations of addition and subtraction, as necessary.
   a. $P(z > 1.06)$ b. $P(z > -1.06)$ c. $P(z < 2.15)$
   d. $P(z < -2.15)$ e. $P(-1.06 < z < 4.00)$ f. $P(1.06 < z < 4.00)$

SELF-CORRECTING EXERCISES FOR CHAPTER 6

1. Find the area under the normal curve that lies between the mean ($z = 0$) and each of the following z scores.
   a. $z = 1.0$ b. $z = 0.58$
   c. $z = 1.68$ d. $z = 3.23$

2. Find the area under the normal curve that lies between the following pairs of z scores.
   a. $z = 0$ and $z = -1.23$ b. $z = 0$ and $z = -0.58$
   c. $z = 0$ and $z = -3.28$ d. $z = 0$ and $z = -2.11$

3. Find the probability that a piece of data picked at random from a normal population will have a z score between each of the following pairs of values.
   a. $z = 0$ and $z = 2.17$ b. $z = 0$ and $z = -1.78$
   c. $z = 0$ and $z = 0.42$ d. $z = 0$ and $z = -0.83$

4. Find the proportion of the area under the normal curve between the following pairs of z scores.
   a. $z = 0$ and 3.50   b. $z = 0$ and -2.11

SELF-CORRECTING EXERCISES

5. Find the area under the normal curve that lies between the following pairs of z-values.
   a. $z = -1.0$ and $z = +2.50$
   b. $z = -1.53$ and $z = +1.87$
   c. $z = -1.50$ and $z = +1.50$
   d. $z = -2.31$ and $z = +3.12$
   e. $z = +1.00$ and $z = +3.20$

6. Find the probability that a piece of data picked at random from a normal population will have a z score between each of the following pairs of values.
   a. $z = -1.50$ and $z = +3.10$
   b. $z = +1.50$ and $z = +3.10$
   c. $z = -1.50$ and $z = -3.10$
   d. $z = -3.10$ and $z = +1.50$

7. Find the proportion of the area under the normal curve that is between:
   a. $z = -1.2$ and $z = +2.4$.
   b. $z = +1.2$ and $z = +2.4$.

8. Find the area under the normal curve that is:
   a. to the right of $z = 0$.
   b. to the right of $z = +1.00$.
   c. to the right of $z = +2.30$.
   d. to the right of $z = -0.85$.
   e. to the left of $z = +1.34$.
   f. to the left of $z = -2.10$.

9. Find the probability that a piece of data selected at random from a normal population has a standard score z that is:
   a. less than +2.56.
   b. greater than +2.56.
   c. greater than -1.52.
   d. less than -2.50.
   e. greater than 1.00.

10. Find the proportion of the area under the normal curve that is:
    a. to the right of $z = 0.0$.
    b. to the right of $z = 0.58$.
    c. to the left of $z = 0.58$.

11. Find each of the following.
    a. $P(0 < z < 0.95)$
    b. $P(-1.25 < z < 0)$
    c. $P(-1.50 < z < +1.50)$
    d. $P(-2.25 < z < +3.25)$
    e. $P(z < +3.25)$
    f. $P(z > -2.40)$

12. Find z_0 such that each of the following statements is true.
    a. The area between $z = 0$ and z_0 is 0.4131.
    b. The area between $z = 0$ and $-z_0$ is 0.4925.
    c. The probability that a randomly selected value of a normal variable has a z score less than z_0 is 0.8554.
    d. The area between $-z_0$ and $+z_0$ is 0.7540.
    e. The area to the right of z_0 is 0.1867.
    f. The probability that z is less than z_0 is 0.9015.

13. Given that x is normally distributed with a mean of 100 and a standard deviation of 10, find each of the following.
    a. The area under the curve between $x = 100$ and $x = 110$.
    b. The area under the curve between $x = 85$ and $x = 112$.
    c. The area to the right of $x = 100$.
    d. The area to the right of $x = 115$.

e. The area between $x = 105$ and $x = 130$.
f. The probability that a randomly selected x is between $x = 75$ and $x = 135$.
g. The probability that a randomly selected x is greater than 110.
h. The proportion of the population that is smaller than 120.

14. Given that x is normally distributed with a mean of 9.3 and a standard deviation of 2.1, find each of the following.
    a. The area under the curve between 8.0 and 10.0.
    b. The area under the curve between 10.0 and 14.5.
    c. The probability that a randomly selected value of x is between 5.0 and 8.0.
    d. The probability that a randomly selected value is greater than 7.5.
    e. The proportion of the population that is greater than 15.0.
    f. The proportion of the population that is less than 12.0.

15. State three interpretations of the statement: "$P(10 < x < 25) = 0.85$," where x is a normally distributed variable.

16. a. Find the probability that the value of x is between 23.2 and 35.7 when x is a randomly selected value taken from a normal population with a mean of 25.0 and a standard deviation of 5.0.
    b. Find $P(56.2 < x < 92.1)$ when x is randomly selected from a normal population with a mean of 48.5 and a standard deviation of 14.6.
    c. Find $P(x > 134.5)$ when x is randomly selected from a normal population with a mean of 128.9 and a standard deviation of 6.2.

17. If the heights of 5,000 students are normally distributed with a mean of 68.5 inches and a standard deviation of 3.2 inches, find each of the following.
    a. The proportion of students who are shorter than 62 inches.
    b. The probability that a person picked at random from the group is taller than 72 inches.
    c. The number of students who are shorter than 62 inches.

18. In a league of 100 bowlers, the bowling averages are normally distributed with a mean of 155.6 and a standard deviation of 7.9.
    a. What percentage of the bowlers in this league have an average below 150?
    b. If a bowler is selected at random from this league, what is the probability that his average is between 150 and 170?
    c. How many of the bowlers have an average above 170?

19. If the weights of a manufactured item are normally distributed with a mean weight of 7 pounds 2.4 ounces (114.4 oz.) and a standard deviation of 1 pound 1.2 ounces (17.2 oz.), find each of the following.
    a. The proportion of weights that are more than 8 pounds.
    b. The proportion of weights that are less than 5 pounds.
    c. The number of items that can be expected to weigh between 7 pounds and 8 pounds in a sample of 500 such items.

SELF-CORRECTING EXERCISES

20. Wine is bottled mechanically by a machine that fills gallon bottles with an average of exactly one gallon of wine each. The machine's controls allow the amount put into each bottle to vary about the mean with a standard deviation of 0.015 gallon. Suppose the amount put into each bottle is a normally distributed variable.
   a. What percentage of the bottles will have less than one gallon in them?
   b. Suppose the machine is to be readjusted (change mean amount dispensed per bottle only; standard deviation is fixed) so that only 10 percent of the bottles have less than one gallon in them. To what value must the mean be changed for this to occur?

SAMPLE VARIABILITY

7

LESSON 7-1: THE CENTRAL LIMIT THEOREM, THEORETICALLY

The central limit theorem claims that the sampling distribution of sample means ($\bar{x}$'s) will display three properties: (1) the mean of the $\bar{x}$'s ($\mu_{\bar{x}}$) will be equal to the mean of the population of individual x's (μ_x), (2) the standard deviation of the $\bar{x}$'s ($\sigma_{\bar{x}}$) will be equal to the standard deviation of the population divided by the square root of the size of each sample ($\sigma_x/\sqrt{n}$), and (3) the $\bar{x}$'s will be approximately normally distributed.

To test this claim theoretically (by using expected probabilities, and so on), let's consider a population formed by the single-digit numbers 1, 2, 3, where each number occurs with equal frequency; that is, $P(1) = 1/3, P(2) = 1/3$, and $P(3) = 1/3$.

Our population has a rectangular distribution shown by the histogram (Figure 7-1). The mean and standard deviation of x within our population are $\mu = 2.0$ and $\sigma = 0.82(\sigma_x)$. The calculation of these values is shown in Table 7-1.

Figure 7-1

Table 7-1

x	$P(x)$	$x \cdot P(x)$	$x^2 \cdot P(x)$
1	1/3	1/3	1/3
2	1/3	2/3	4/3
3	1/3	3/3	9/3
	3/3 (ck)	6/3 = 2	14/3

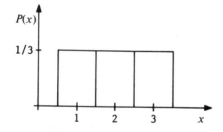

$\mu = \Sigma x \cdot P(x) = 2.0$

$\sigma = \sqrt{\Sigma x^2 P(x) - [\Sigma x P(x)]^2}$

$= \sqrt{14/3 - (2)^2} = \sqrt{2/3}$

CENTRAL LIMIT THEOREM, THEORETICALLY Lesson 7-1

Let's also consider random samples of size 3 drawn from this population. Table 7-2 shows a list of all the various samples of size 3 that could be drawn from our population. Each sample is represented by an ordered three-digit number, where the first digit represents the first digit drawn, second digit represents the second digit drawn, and so on.

Table 7-2

111	211	311
112	212	312
113	213	313
121	221	321
122	222	322
123	223	323
131	231	331
132	232	332
133	233	333

Each one of these twenty-seven samples is equally likely to occur, since the probabilities of the individual digits are each 1/3 and remain fixed. Therefore the probability assigned to the likeliness of each individual sample is 1/27. Table 7-3 shows these samples with the mean of each sample in parentheses.

Table 7-3

111 (3/3)	211 (4/3)	311 (5/3)
112 (4/3)	212 (5/3)	312 (6/3)
113 (5/3)	213 (6/3)	313 (7/3)
121 (4/3)	221 (5/3)	321 (6/3)
122 (5/3)	222 (6/3)	322 (7/3)
123 (6/3)	223 (7/3)	323 (8/3)
131 (5/3)	231 (6/3)	331 (7/3)
132 (6/3)	232 (7/3)	332 (8/3)
133 (7/3)	233 (8/3)	333 (9/3)

Since many of the values of sample means repeat and we theorize that each sample is equally likely, we will show the results listed above as a probability distribution (Table 7-4). (We did not take any samples; we only theorized what would happen if we were to take samples — thus we use probabilities.)

Table 7-4

Sample Mean $\bar{x}$	3/3	4/3	5/3	6/3	7/3	8/3	9/3	
Expected Probability $P(\bar{x})$	1/27	3/27	6/27	7/27	6/27	3/27	1/27	27/27 = 1 (ck)

SAMPLE VARIABILITY Chapter 7

The central limit theorem describes three properties. $\mu_{\bar{x}}$, $\sigma_{\bar{x}}$, and the distribution. First let's calculate the mean and standard deviation of the $\bar{x}$ distribution, $\mu_{\bar{x}}$ and $\sigma_{\bar{x}}$, respectively (Table 7-5).

Table 7-5

$\bar{x}$	$P(\bar{x})$	$\bar{x} \cdot P(\bar{x})$	$\bar{x}^2 P(\bar{x})$
3/3	1/27	3/81	9/243
4/3	3/27	12/81	48/243
5/3	6/27	30/81	150/243
6/3	7/27	42/81	252/243
7/3	6/27	42/81	294/243
8/3	3/27	24/81	192/243
9/3	1/27	9/81	81/243
	27/27 = 1 (ck)	162/81 = 2.0	1026/243 = 38/9

$$\mu_{\bar{x}} = \Sigma \bar{x} \cdot P(\bar{x}) = 162/81 = \underline{2.0}$$
$$\sigma_{\bar{x}} = \sqrt{\Sigma \bar{x}^2 \cdot P(\bar{x}) - [\Sigma \bar{x} \cdot P(\bar{x})]^2} = \sqrt{38/9 - (2)^2} = \underline{\sqrt{2/9}}$$

The theoretical sampling distribution of $\bar{x}$'s is shown in the following histogram (Figure 7-2).

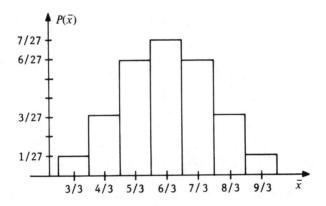

Figure 7-2

Now to compare our findings. (1) We found $\mu_{\bar{x}} = 2.0$, which is exactly the same value as $\mu_x = 2.0$, found earlier. (2) We found $\sigma_{\bar{x}} = \sqrt{2/9}$; and earlier we found $\sigma_x = \sqrt{2/3}$. If we divide the $\sqrt{2/3}$ (σ_x) by the square root of n, the size of any one sample, we obtain:

$$\frac{\sqrt{2/3}}{\sqrt{3}} = \sqrt{\frac{2/3}{3}} = \sqrt{2/9}$$

And this is the same value that was found as $\sigma_{\bar{x}}$. (3) Inspection of the histogram shows that $\bar{x}$ is approximately normally distributed.

Therefore we have seen that all three properties claimed by the central limit theorem have occurred in this situation. This does not prove the central limit theorem, but it should help convince you of the possibility that it is correct.

EXERCISES 7-1

1. Consider a population of {0, 2, 4, 6}, where each digit is equally likely and the distribution of means of samples of size 2 is drawn randomly from this population. Verify the three aspects of the central limit theorem by finishing the following.
    a. Information about population: mean and standard deviation.
    b. List all possible samples and calculate the mean of each sample.
    c. Form a probability distribution of the possible sample means.
    d. Draw a histogram and calculate the mean and standard deviation of this theoretical sampling distribution of sample means.
    e. Check the three aspects of the Central Limit Theorem.

LESSON 7-2: THE CENTRAL LIMIT THEOREM, EMPIRICALLY

Does the central limit theorem hold up in actual practice? That is the question we want to examine in this lesson. In Lesson 7-1 we saw the claims of the central limit theorem hold up in a theoretical (probabilistic) setting. Now we would like to see if they hold when the samples are actually drawn from a given population. The population to be considered for this study is the set of digits {0, 1, 2, 3, 4}. Each is equally likely.

The mean of our population is 2.0 and the standard deviation is 1.414, as shown in the following calculations (Table 7-6). The rectangular distribution of our population in the histogram in Figure 7-3.

Table 7-6

x	$P(x)$	$xP(x)$	$x^2 P(x)$
0	1/5	0	0
1	1/5	1/5	1/5
2	1/5	2/5	4/5
3	1/5	3/5	9/5
4	1/5	4/5	16/5
	5/5 = 1.0	10/5 = 2.0	30/5 = 6.0

$$\mu_x = \underline{2.0} \text{ and } \sigma_x = \sqrt{6.0 - (2.0)^2} = \sqrt{2} = \underline{1.4}$$

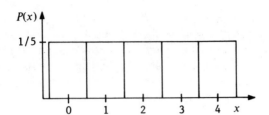

Figure 7-3

We will draw 50 samples of size 2 from this population. Below you will find the list of the 50 samples and their means (Table 7-7).

Table 7-7

Sample	$\bar{x}$	Sample	$\bar{x}$	Sample	$\bar{x}$	Sample	$\bar{x}$	Sample	$\bar{x}$
14	2.5	43	3.5	42	3.0	24	3.0	44	4.0
23	2.5	14	2.5	21	1.5	02	1.0	24	3.0
33	3.0	22	2.0	22	2.0	33	3.0	14	2.5
11	1.0	00	0.0	01	0.5	24	3.0	31	2.0
12	1.5	22	2.0	41	2.5	33	3.0	03	1.5
43	3.5	10	0.5	34	3.5	30	1.5	23	2.5
21	1.5	32	2.5	34	3.5	43	3.5	20	1.0
31	2.0	11	1.0	12	1.5	34	3.5	10	0.5
13	2.0	14	2.5	00	0.0	41	2.5	23	2.5
04	2.0	11	1.0	14	2.5	01	0.5	13	2.0

By constructing an ungrouped frequency distribution of these 50 sample means ($\bar{x}$'s) (Table 7-8), we can draw a histogram (Figure 7-4) and calculate the mean and standard deviation for the observed sampling distribution of the sample means. (These are observed samples, therefore "frequencies" are used instead of probabilities.)

Table 7-8

$\bar{x}$	f	$\bar{x}f$	$\bar{x}^2 f$
0.0	2	0.0	0.00
0.5	4	2.0	1.00
1.0	5	5.0	5.00
1.5	6	9.0	13.50
2.0	8	16.0	32.00
2.5	11	27.5	68.75
3.0	7	21.0	63.00
3.5	6	21.0	73.50
4.0	1	4.0	16.00
4.5	0	0.0	0.00
	50	105.5	272.75

CENTRAL LIMIT THEOREM, EMPIRICALLY   Lesson 7-2

Evidence A. The observed mean of sample means:

$$\bar{\bar{x}} = \frac{\Sigma \bar{x} f}{\Sigma f} = \frac{105.5}{50} = 2.11$$

The standard deviation of sample means:

$$s_{\bar{x}} = \sqrt{\frac{(\Sigma f)(\Sigma \bar{x}^2 f) - (\Sigma \bar{x} f)^2}{(\Sigma f)(\Sigma f - 1)}}$$

$$= \sqrt{\frac{(50)(272.75) - (105.5)(105.5)}{(50)(49)}}$$

$$= \sqrt{1.0233} = 1.0115$$

The observed mean ($\bar{\bar{x}}$) and standard deviation ($s_{\bar{x}}$) represent an estimation for the mean and standard deviation of the theoretical sampling distribution ($\mu_{\bar{x}}$ and $\sigma_{\bar{x}}$, respectively). Therefore we should expect to find $\bar{\bar{x}}$ to be approximately equal to μ_x, the population mean. The 2.11 is reasonably close to the population mean of 2.0 and the $s_{\bar{x}} = 1.0115$ is very close to the value of $\sigma/\sqrt{n}$, which is 1.0 ($\sigma_x/\sqrt{n} = \sqrt{2}/\sqrt{2} = 1.0$).

Evidence B. n is the size of one sample. Each sample is made up of two observations; therefore $n = 2$. The histogram showing the distribution of the 50 observed sample means suggests a normal distribution (Figure 7-4).

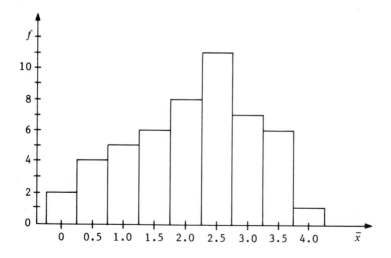

Figure 7-4

Therefore, we must conclude that all three claims made by the central limit theorem are reasonable in view of our evidence. This evidence does not prove the correctness of the theorem, but it seems that all the evidence found thus far does indeed support its claims.

EXERCISES 7-2

1. Consider a population of three equally likely digits, {0, 2, 4}. Take 50 samples of size 2 from this population and form a sample for the sampling distribution (an observed frequency distribution). (You can use dice or the random-number table to obtain your samples.) Compare the mean, standard deviation, and distribution of these 50 $\bar{x}$'s to the population to check the validity of the central limit theorem.

LESSON 7-3: APPLICATION OF THE CENTRAL LIMIT THEOREM

In Lessons 7-1 and 7-2 we studied the truth of the central limit theorem. By now you should be convinced that it is in fact a true statement. But, you might ask, how is this theorem going to be of any help to us in statistics?

The central limit theorem tells us about the pattern of variability for means of samples of size n; that is, it tells us how sample means will be distributed in reference to the mean and standard deviation of the population from which the sample was taken. Now let's assume that we have some population for which we know the mean μ and the standard deviation σ, say, $\mu = 50$ and $\sigma = 10$. If we were to draw one sample of size 4 and calculate its mean ($\bar{x}$), within what bounds would we expect to find this value of $\bar{x}$? The central limit theorem tells us that all such $\bar{x}$'s for samples of size 4 are distributed about a mean of 50 ($\mu_{\bar{x}} = \mu$, which is 50) with a standard deviation of 5 ($\sigma_{\bar{x}} = \sigma/\sqrt{n} = 10/\sqrt{4} = 5$, since $\sigma = 10$ and $n = 4$). We also know that the distribution of the $\bar{x}$'s is approximately normally distributed regardless of the shape of the original population distribution. Therefore, the distribution of all possible $\bar{x}$'s for samples of size 4 can be expected to look like Figure 7-5.

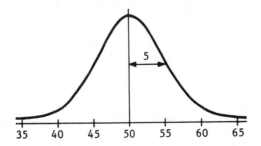

Figure 7-5

APPLICATION OF CENTRAL LIMIT THEOREM   Lesson 7-3

By using the information available to us from the use of the normal distribution, we can expect our one observed sample mean to have a value between 45 and 55, with a probability of 0.6826 [$P(45 < \bar{x} < 55) = P(-1 < z < +1) = 2(0.3413) = 0.6826$]. With 0.9544 probability, we can expect our sample mean to be between 40 and 60 [$P(40 < \bar{x} < 60) = P(-2.0 < z < +2.0) = 2(0.4772) = 0.9544$].

In practice, we will take only one sample from our population. We would not expect another sample to yield the same value of $\bar{x}$, nor would we expect the sample mean to be exactly equal to the population mean. But with the aid of the central limit theorem, we will be able to make probability statements about where we expect this one value, our sample mean, to be. In Chapter 8 we will discuss how to proceed when our interest is shifted from the population mean to the sample mean, and inferences are made using the sample mean as a basis.

Illustration What is the probability that the mean of a sample will be within 2 units of the mean of a given population, where $\mu = 50$ and $\sigma = 10$, if the sample contains 25 pieces of data?

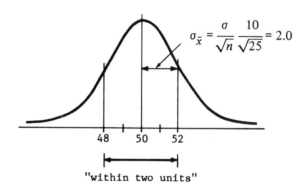

Figure 7-6   "within two units"

Solution

$$P(48 < \bar{x} < 52) = P\left(\frac{48 - 50}{2} < z < \frac{52 - 50}{2}\right)$$

$$= P(-1.0 < z < +1.0)$$

$$= 2(.3413) = \underline{0.68}$$

Illustration What is the probability that the mean of the sample in the above illustration is greater than 51?

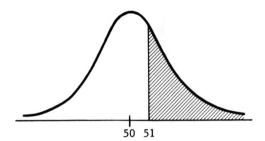

Figure 7-7

Solution

$$P(\bar{x} > 51) = P(z > 0.50) = 0.5000 - 0.1915 = \underline{0.3085}$$

$$[z = (\bar{x} - \mu)/(\sigma/\sqrt{n}) = (51 - 50)/(10/\sqrt{25}) = 1/2 = 0.50.]$$

EXERCISES 7-3

1. A sample of size 16 is drawn from a population where $\mu = 26$ and $\sigma = 4.5$.
   a. Find the probability that the sample mean is between 24 and 28.
   b. Find the probability that the sample mean is less than 27.
   c. Find the probability that the sample mean is between 25 and 30.

2. A sample of size 150 is drawn from a population whose mean is 62.4 and whose standard deviation is 10.8. Find the probability that $\bar{x}$ will be:
   a. within 0.75 units of the mean.
   b. less than 63.0.

3. A sample is to be drawn from a population whose mean is 30 and whose standard deviation is 6. What size sample must be drawn for a 90 percent chance that $\bar{x}$ will be between 29 and 31?

SELF-CORRECTING EXERCISES FOR CHAPTER 7

1. Use the table of random digits (Table 1, page 566 in the textbook) as your population.
   a. Draw a random sample of size 5 from this population and record it as sample 2 on the chart below. Calculate Σx and $\bar{x}$ for your sample. (Sample 1 has already been done.)

SELF-CORRECTING EXERCISES

Sample		Σx	$\bar{x}$
(1)	9, 8, 4, 1, 6	28	5.6
(2)	·		
	·		
	·		

   b. Repeat the above sampling until you have a total of 20 different samples on the list.
   c. Form a grouped frequency distribution with the list of 20 $\bar{x}$'s from above using class boundaries of 0.5, 1.5, 2.5, 3.5, and so on.
   d. Draw the histogram showing the grouped frequency distribution found in part c.

2. a. Using the same set of 20 samples found in Exercise 1a, add a column to your chart and find the median for each sample. (The median for the first sample is 6.)
   b. Form a grouped frequency distribution of the above 20 medians and draw the histogram showing this distribution. Use class boundaries of 0.75, 2.25, 3.75, 5.25, 6.75, 8.25.

3. a. Using the same set of 20 samples found in Exercise 1a, add another column and find the midrange for each sample.
   b. Form a grouped frequency distribution of the above 20 midranges and draw the histogram showing this distribution. Use class boundaries of 0.75, 2.25, 3.75, 5.25, 6.75, 8.25.

4. a. Using the 20 samples found in Exercise 1, find the range of each sample.
   b. Form an ungrouped frequency distribution of these 20 ranges and draw a histogram to show the distribution.

5. a. Using the 20 samples found in Exercise 1, find the standard deviation of each sample.
   b. Form a grouped frequency distribution of these 20 standard deviations and draw the histogram of it. Use class boundaries of 2.0, 2.5, 3.0, and so on.

6. a. Calculate the mean and the standard deviation of the 20 $\bar{x}$'s in Exercise 1, $\bar{\bar{x}}$ and $s_{\bar{x}}$, respectively.
   b. The distribution of random single-digit numbers has a mean of 4.5 and a standard deviation of 2.87. According to the central limit theorem, the mean value, $\mu_{\bar{x}}$, for all such samples of size 5 should be what value? How does your observed value, $\bar{\bar{x}}$, compare with the $\mu_{\bar{x}}$ claimed by the CLT?
   c. According to the CLT, the standard error of sample means, $\sigma_{\bar{x}}$, for all such samples of size 5 should be what value? How does your observed value, $s_{\bar{x}}$, compare to the $\sigma_{\bar{x}}$ claimed by the CLT?

SAMPLE VARIABILITY Chapter 7

7. A random sample of size 16 is to be taken from a population of mean 60 and standard deviation 12.
   a. According to the central limit theorem, what will be the mean value of all such samples of size 16 taken from this population?
   b. According to the CLT, what will be the value of the standard deviation of the means of all such samples of size 16?
   c. According to the CLT, how will all the sample means be distributed?

8. A random sample of size 100 is to be taken from a population whose distribution has a mean value of 78.0 and a standard deviation of 12.5. The sampling distribution of the means of all such samples taken from this population will be distributed as described by three basic properties.
   a. The mean of this distribution of $\bar{x}$'s will be ___?___.
   b. The standard deviation of this distribution will be ___?___.
   c. The shape of this distribution of $\bar{x}$'s will be ___?___.

9. Apply the information provided by the central limit theorem and find the following probabilities concerning the mean of a random sample of size 25 drawn from a population with mean of 48 and standard deviation of 8.
   a. $P(\bar{x} > 48)$
   b. $P(48 < \bar{x} < 51)$
   c. $P(48 < \bar{x} < 53)$
   d. $P(44 < \bar{x} < 48)$
   e. $P(44 < \bar{x} < 52)$
   f. $P(\bar{x} < 52)$

10. Assume that the heights of college students are distributed about a mean μ of 68.5 inches with a standard deviation σ of 3.5 inches. A random sample of 25 students is to be selected. Find the probability that the sample mean is:
    a. greater than 70 inches.
    b. between 67 and 71 inches.
    c. greater than 68 inches.
    d. less than 67.5 inches.

11. If the daily wages earned in a particular industry are normally distributed with a mean daily wage of $64.50 and a standard deviation of $11.80, find the probability that the mean of a random sample of size 16 is:
    a. greater than $60.
    b. between $65 and $70.
    c. greater than $68.
    d. less than $72.
    e. between $62.50 and $66.50.
    f. Less than $65.50.

12. A manufacturer of light bulbs claims that his light bulbs have a mean life of 1,250 hours with a standard deviation of 75 hours. A random sample of 25 such bulbs is selected to be tested.
    a. Find the probability that the sample mean of these 25 bulbs is greater than 1,250 hours.
    b. Find the probability that the sample mean is greater than 1,225 hours.
    c. What is the probability that the sample mean is no more than 1,300?

INTRODUCTION TO STATISTICAL INFERENCES

8

LESSON 8-1: THE NATURE OF HYPOTHESIS TESTING

Many of the kits sold to schools for statistical sampling include a mechanical "coin tosser" (Figure 8-1). This machine is for "tossing" a coin, with the human element of control removed. (Many people think they can somewhat control the result of flipping a coin with their thumb; thus this machine removes that bias.) But we might question whether the coin tosser really is fair — perhaps the side of the coin placed up when the machine is set determines the side that will be up after the coin has landed. Now we must test the coin tosser if we are to decide whether we agree or disagree with the claim that it is unbiased. This testing procedure (planning, experimenting, and making a decision) is known as a hypothesis test.

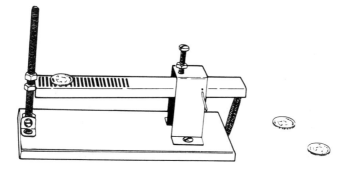

Figure 8-1

A hypothesis is a claim about some aspect of a situation. The hypothesis in our illustration would be the claim about the fairness (bias or lack of bias) of the machine as a tosser of coins. To test this claim, we will formulate a set of hypotheses about some parameter of the population associated with the variable used to measure the fairness of the coin tosser. This set of hypotheses will always include exactly two statements, one of which we will conclude to be true.

The concept of "fairness" that seems appropriate for our illustration could be measured by counting the number of times (x) that the coin lands with the same side up as was placed up on the coin tosser. n would be the number of test flips that were made, and $n - x$ would be the number of times the coin landed upside down from its initial position. If the coin tosser is fair, then we would expect to find x and $n - x$ to be approximately equal $[x \approx n/2; (n - x) \approx n/2]$.

To test a hypothesis, follow these five basic steps: (1 and 2) identify the two opposing possibilities (fair or not fair, for our illustration), (3) determine the testing ground rules, (4) do the actual experimental testing, and (5) draw your conclusions. Let's now look at each step in terms of our illustration.

Step 1. We have already indicated that the two possible statements of opposing views are (a) the machine does toss coins fairly or (b) the machine does not toss coins fairly. The population parameter here is the probability that the coin lands with the same side up as its position when placed on the tosser. "Fair" would be implied when the true probability is 0.5. "Not fair" would be implied when the true probability is different from 0.5. (If the value of p is between 0 and 0.5, the machine gives us the opposite side most of the time; whereas if p is between 0.5 and 1.0, the machine is giving us the same side most of the time.) The two statements above, (a) and (b), are used as the null hypothesis and the alternative hypothesis, described in the textbook on pages 293-301. The "equal" statement becomes the null hypothesis — the claim that is "on trial" during the test. The null hypothesis, then, states that the probability that a coin lands with the same side up as in its initial position is 0.5 ($H_0: p = 0.5$). The interpretation of the null hypothesis is "fair." The alternative is that the probability is not 0.5 ($H_a: p \neq 0.5$), which is interpreted as "not fair."

Step 2. Next we must determine the "ground rules" for the testing procedure (test criteria). (In this lesson we should like to present the general procedures rather than the specific details of hypothesis testing; some of the ideas outlined will become modified as we study the various types of hypothesis testing, but the basic format and rationale will remain the same.) The test statistic to be used is x, the number of times that the coin lands with the same side up. We must also determine the number of flips to be used in testing our machine. Let's decide to flip a coin twelve times. This may be too few to inspire confidence in our results, but it will serve as an illustration for the entire procedure. Bear in mind that $n = 12$ and that x is the number of "same" results obtained. x could, at the end of twelve flips, be any integer from 0 to 12. As previously stated, values of x around 6 will mean "fair," values of x farther from 6 will be interpreted as "not fair." See Figure 8-2.

Now we need to locate the "decision points" (critical values), the points at which our interpretation of the observed value of x changes. Recall that we said

Figure 8-2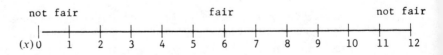

NATURE OF HYPOTHESIS TESTING   Lesson 8-1

$x = 6$ would be called "fair" and $x = 0$ or $x = 12$ would be called "not fair." What about all the other possible values for x? Suppose that after some thought we decide to consider $x = 4, 5, 6, 7,$ and 8 as fair and $x = 0, 1, 2, 3, 9, 10, 11,$ and 12 as not fair (Figure 8-3). These values then would be the "test criteria," the game rules, if you wish. They must be established prior to the experiment so the results won't bias your decision on the location of these decision points.

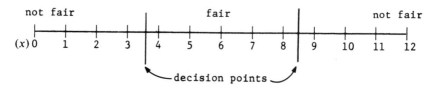

Figure 8-3

Steps 3 and 4. Now we are ready to do the experiment and obtain our test statistic. Upon completion of the experiment we will compare the test statistic (x for this illustration) to the test criteria, make a decision about the null hypothesis, and draw a conclusion.

Recall that we stated earlier that the null hypothesis was on trial. Basically the hypothesis-testing procedure works like the courts of law: "The accused is innocent until proved guilty." In our hypothesis test, "The null hypothesis is true until proved false." Well, almost. We proceed under the assumption that the parameter value claimed in the null hypothesis is correct and then make a decision to disagree or not to disagree with the value claimed. We *cannot* prove or disprove a null hypothesis by our procedure. (The reasons for this are explained in the next lesson.)

EXERCISES 8-1

1. A friend claims, "I can flip a penny with my thumb and get a head most of the time." Use x as the number of heads that result in twelve test flips and set up the hypothesis (null and alternative) and the test criteria that you think would be reasonable to use in testing the friend's claim. Describe the five basic aspects of the hypothesis-testing procedure.

2. A member of the varsity basketball team claims to be able to make at least 80 percent of the free-throw shots that he takes. Use y as the variable number of baskets made in a test consisting of fourteen attempts. Describe the hypotheses and test criteria that you would use to check on his claim. Describe in detail.

3. A student at our college claimed that the average hourly wage earned by the working student is $3.50. Using w as the variable hourly wage earned by each working student in a sample of twenty students, describe the five aspects of the hypothesis-testing procedure.

LESSON 8-2: ALPHA AND BETA

Alpha and beta are the probabilities of making the two kinds of errors that exist in hypothesis testing. Alpha, α, is the probability of rejecting H_0 when it is a true statement (type I error). Beta, β, is the probability of failing to reject H_0 when it is a false setatemtn (type II error). Let's inspect these two errors and their probabilities in the hypothesis test illustration discussed in Lesson 8-1.

Note: Recall that when testing a hypothesis, we assume the null hypothesis to be correct and reach one of two possible decisions: (1) "reject H_0," meaning that the evidence found during experimentation suggests that the null hypothesis is false, and (2) "fail to reject H_0," meaning that the evidence does not contradict the claim of the null hypothesis. Notice that we find evidence that "contradicts" or "does not contradict" the null hypothesis. Remember that it is H_0 that is on trial, not H_a. The frame of reference is important.

We perform the hypothesis test under the assumption that the null hypothesis (always contains the equal relationship) is true until evidence dictates that we reject it. This may or may not be a correct assumption, since the claim put forth in the null hypothesis could be either true or false. This is where the two errors in decision making come into view. If H_0 is actually true, we could find evidence that agrees or disagrees with its claim. If the evidence found disagrees, then we will reject H_0, claiming it to be false when in fact it is correct. This is a type I error, and the probability of its occurrence in a given situation is called "alpha."

Let's return to our illustration using the coin tosser and the information given in Lesson 8-1:

1. $H_0: p = P(\text{"same"}) = 0.5$ (fair)
   $H_a: p \neq 0.5$ (not fair)
2. Test criteria (Figure 8-4):

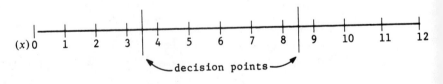

Figure 8-4

As you can see from Table 8-1, the probability that the observed value of x falls in the rejection region is 0.146 even if the coin tosser does operate perfectly fairly.

ALPHA AND BETA   Lesson 8-2

Table 8-1

n	x	Binomial probability when $p = 0.5$*
12	0	0.0+
	1	0.003
	2	0.016
	3	0.054
	9	0.054
	10	0.016
	11	0.003
	12	0.0+
		$\alpha = 0.146$

$\alpha = P(\text{type I error})$
$= P(\text{reject } H_0 \text{ when } H_0 \text{ is actually true})$

*See Table 4, p. 570 in textbook.

Thus, as we designed our test criteria in Lesson 8-1, we would commit the type I error with a probability of almost 15 percent.

In practice we will not let the decision points dictate the probability of the type I error (α), but instead will develop and use a system where the level of (value of) alpha can be tolerated is determined by the seriousness of the type I error, and then let α in turn dictate the location of the decision points. As the remainder of Chapter 9 unfolds, you will see that α is controlled directly; that is, we will set its level and work from there. At present the same is not true for β, the probability of the type II error.

The type II error is made when the null hypothesis is false to start with and we end up making the decision that it appears to be true. The probability of this error is called β. Let's use our same illustration and investigate β, assuming that the coin tosser is actually quite controlled and the true value of p, the probability that the same side landed up as was placed up on the tosser, is 0.8 (Table 8-2).

Table 8-2

n	x	Binomial probability when $p = 0.8$*
12	4	0.001
	5	0.003
	6	0.016
	7	0.053
	8	0.133
		$\beta = 0.206$

$\beta = P(\text{type of II error})$
$= P(\text{fail to reject } H_0 \text{ when } H_0 \text{ is actually false})$

Thus, we will obtain a value of x that belongs to the "fail to reject region" of the test criteria, when p is truly equal to 0.8, almost 21 percent of the time.

Note: You cannot commit both errors at the same time, and their probabilities are not directly related.

In practice, the level of β should be dictated by the seriousness of its occurrence. Then α and β could be used to determine the size of the sample that would be required for a given test. Because we have not studied enough statistics at this point to tackle the beta concept, we will choose a sample size somewhat arbitrarily for now. (The effect and use of beta are beyond the scope of an elementary textbook.)

EXERCISES 8-2

1. Refer to Exercise 1 in lesson 8-1.
    a. Carefully describe the meaning of type I and type II errors.
    b. Calculate the value of α that corresponds to the test as described.
    c. Find the value of β that exists if the friend does have control over the results, that is, if p is really equal to 0.7.

2. a. Find the value assigned to α in the hypothesis test for the varsity basketball player in Exercise 2 of Lesson 8-1.
    b. Calculate the value of β, if he really makes only one-half of his free throws.

3. a. Find the value of α in the hypothesis test for the coin tosser in Exercise 1 of Lesson 8-1, if he really has 80 percent control.
    b. Find the value of β in the same hypothesis test if he really has no control at all; that is, if $p = 0.5$.

LESSON 8-3: THE NATURE OF ESTIMATION

When a sample is drawn for making an estimate, the estimate will be one of two kinds: (1) a point estimate or (2) an interval estimate. The point estimate for a population parameter is obtained by calculating the corresponding sample statistic. For example, if we want to estimate the mean value of a particular population, we simply take a sample from this population and report the sample mean as the point estimate.

The interval estimate (confidence-interval estimate) for any population parameter starts out the same way. A sample is obtained from the population and the resulting sample statistic is used as the basis for constructing the interval estimate. The level of confidence, sometimes called confidence coefficient, is the percentage of the time we can expect the true value of the population parameter to fall within the calculated interval. This confidence level is identified by $1 - \alpha$. The use of α here is consistent with its definition as the probability of a type I

NATURE OF ESTIMATION   Lesson 8-3

error; however, the term "type I error" has no meaning in this particular context. The interval either contains the true value or it does not.

To describe how an interval estimate is made, we will estimate the mean value of a population in a situation where the population standard deviation σ is known. (Other interval estimates follow much the same general procedure, except that the sampling distribution for the statistic involved will change and thus an appropriate adjustment will be required.) By looking at an illustration where we have σ as a known quantity, the central limit theorem will provide us with all the necessary information to construct the interval estimate. Let's assume that we want to estimate the mean of a very large, normally distributed population whose standard deviation is known to be 10 units. From this population we obtain one random sample of size 25, whose mean, $\bar{x}$, is 64.2 units.

Now before we use this 64.2, let's look at the sampling distribution of all means of samples of size 25, of which 64.2 is one, and see what the central limit theorem tells us. Recall the three basic facts: (1) the mean of the $\bar{x}$'s, $\mu_{\bar{x}}$, is equal to the mean of the population, μ (it is unknown and we would like to estimate it); (2) the standard deviation of these $\bar{x}$'s is $\sigma_{\bar{x}}$ and equal to $\sigma/\sqrt{n}$, which is $10/\sqrt{25} = 2.0$; and (3) these $\bar{x}$'s are approximately normally distributed. That is, we have observed one value, one $\bar{x}$, from the distribution shown in Figure 8-5.

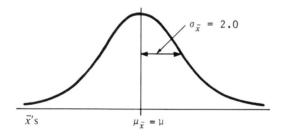

Figure 8-5

We don't know where on this distribution the value of 64.2 lies, but we can make a probability statement about where we could expect it to be found. For example, with 95 percent confidence we could say that the observed $\bar{x}$ was somewhere within the interval from 1.96 standard deviations below the mean ($\mu - 1.96\sigma_{\bar{x}}$) to 1.96 standard deviations above the mean ($\mu + 1.96\sigma_{\bar{x}}$). That is to say, the following statement is true with regard to any one individual sample mean drawn from our population:

$$P(\mu - 1.96\sigma_{\bar{x}} < \bar{x} < \mu + 1.96\sigma_{\bar{x}}) = 0.95$$

This probability statement is a direct result of applying the central limit theorem. Now look carefully at what it states with respect to distance. Notice that $\sigma_{\bar{x}} = 2.0$, and the probability statement can be rewritten to read: $P(\mu - 3.92 < \bar{x} < \mu + 3.92) = 0.95$ (see Figure 8-6). To be within this interval

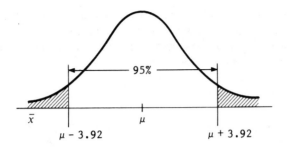

Figure 8-6

means that the distance from μ to $\bar{x}$ is less than 3.92 units. Certainly the distance from $\bar{x}$ to μ is the same as from μ to $\bar{x}$; thus, we could say that any time the distance from μ to $\bar{x}$ is less than 3.92 units, the distance from $\bar{x}$ to μ is also less than 3.92. Therefore the following two probability statements must be equivalent:

$$P(\mu - 3.92 < \bar{x} < \mu + 3.92)$$

and

$$P(\bar{x} - 3.92 < \mu < \bar{x} + 3.92)$$

If this is in fact true, then $P(\bar{x} - 3.92 < \mu < \bar{x} + 3.92)$ has the value of 0.95, and we have made the following interval estimate: With 95 percent confidence we can say that μ is between $\bar{x} - 3.92$ and $\bar{x} + 3.92$. $\bar{x}$ is 64.2; therefore our interval estimate is from 60.28 to 68.12, and we can expect that 95 percent of the intervals constructed in this way will contain the true value of the population mean.

Let's look back now and see how the individual parts of the interval estimation came about. First, $\bar{x}$ is the mean of the sample. $\sigma_{\bar{x}}$ is found by dividing σ, the population standard deviation, by the square root of the sample size, n. The only other value required to calculate the confidence-interval estimate is the number of multiples of $\sigma_{\bar{x}}$ that will be needed to obtain the desired level of confidence. The standard normal score, z, is used for dealing with one sample where σ is known. The value that is needed is $z(\alpha/2)$, the value of z being such that $\alpha/2$ is the measure of the area under the curve to the right of the z score (Figure 8-7).

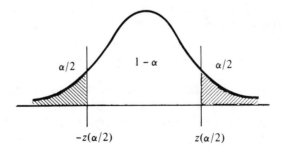

Figure 8-7

SELF-CORRECTING EXERCISES

Figure 8-7 shows $1 - \alpha$ as the level of confidence with $\alpha/2$ as the measure outside the interval on either side. Since the normal distribution is symmetrical about $z = 0$, the two values differ only by the sign; thus $\pm z(\alpha/2)$ is used. The confidence-interval formula 8-2:

$$\bar{x} - z(\alpha/2)(\sigma/\sqrt{n}) < \mu < \bar{x} + z(\alpha/2)(\sigma/\sqrt{n})$$

is often shortened to:

$$\bar{x} \pm z(\alpha/2)(\sigma/\sqrt{n})$$

and the answer is then expressed as an interval.

EXERCISES 8-3

1. A random sample of size 16 is drawn from a normally distributed population whose standard deviation is 12. The sample mean is 19.5.
   a. Find the value of $z(\alpha/2)$ that would be used to construct a 90 percent confidence interval.
   b. Calculate the 90 percent confidence interval for the estimate of the true mean for the above population.

2. Draw a random sample of 36 single-digit numbers from Table 1 in your textbook.
   a. Calculate the sample mean.
   b. Using the fact that the standard deviation of single-digit random numbers is 2.87, calculate the 75 percent confidence interval for the true mean of the single-digit random numbers of Table 1.
   c. Describe the meaning of the 75 percent confidence interval in reference to what you would expect to happen if you were to draw 100 such samples and construct 100 such confidence intervals.

SELF-CORRECTING EXERCISES FOR CHAPTER 8

1. For each of the following situations, state the null and alternative hypotheses.
   a. The mean age of last year's June brides was 20.3 years.
   b. The mean monthly rental fee for second-floor two-bedroom apartments in our city is at least $150.
   c. The mean family size in our state is less than 4.3.
   d. The mean distance to work traveled by the employees of Company X is greater than 7.5 miles.
   e. The mean amount of time per week that housewives spend watching television is 14.5 hours.
   f. The mean amount of money per week that households in our town spend for groceries is at least $42.

2. Suppose we test the hypothesis that the monthly rental fee for a second-floor two-bedroom apartment in our city is at least $150.
   a. Explain the situation if we make the correct decision identified as A.
   b. Explain the situation if correct decision B is reached.
   c. Explain the situation if the type I error is committed.
   d. Explain the situation if the type II error is committed.
   e. Explain the conclusion that would be reached if the decision is "reject H_0."
   f. Explain the conclusion that would be reached if the decision is "fail to reject H_0."

3. The claim, "the mean distance to work traveled by the employees of Company X is greater than 7.5 miles," is tested (parts a through d). Explain the meaning of the situation if:
   a. the correct decision A is reached.
   b. the correct decision B is reached.
   c. the type I error is committed.
   d. the type II error is committed.
   e. If a decision of "fail to reject H_0" is reached, then explain the conclusion.
   f. If a decision of "reject H_0" is reached, then explain the conclusion.

4. Use Table 5 on page 573 of the textbook and determine the critical values of the standard normal z (names and values) shown on the following sketches.

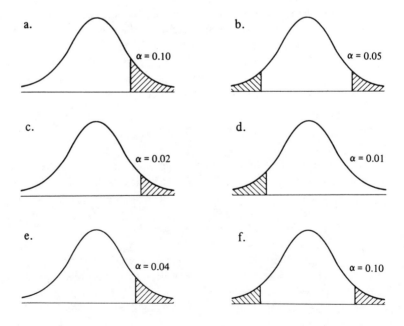

SELF-CORRECTING EXERCISES

g.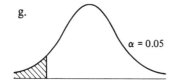
$\alpha = 0.05$

h.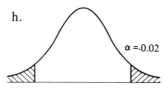
$\alpha = -0.02$

5. Using Table 5 on page 573 in the textbook, find each of the following z values.
   a. $z(0.05)$
   b. $z(0.005)$
   c. $z(0.025)$
   d. $z(0.10)$
   e. $z(0.04)$
   f. $z(0.95)$
   g. $z(0.99)$
   h. $z(0.975)$
   i. $z(0.98)$
   j. $z(0.995)$

6. Determine the text criteria (test statistic, critical value(s), and critical region) that would be used in a hypothesis test of each of the following cases.
   a. H_0: $\mu = 12$ ($\alpha = 0.05$)
      H_a: $\mu \neq 12$
   b. H_0: $\mu = 15.2$ ($\alpha = 0.05$)
      H_a: $\mu > 15.2$
   c. H_0: $\mu = 9.86$ ($\alpha = 0.05$)
      H_a: $\mu < 9.86$
   d. H_0: $\mu = 100$ ($\alpha = 0.02$)
      H_a: $\mu < 100$
   e. H_0: $\mu = 115.5$ ($\alpha = 0.01$)
      H_a: $\mu > 115.5$
   f. H_0: $\mu = 5.38$ ($\alpha = 0.10$)
      H_a: $\mu \neq 5.38$

7. Test the hypothesis that the mean family size in our state is 4.3. A random sample of 20 families resulted in a mean of 5.1. (Use $\sigma = 1.0$ and $\alpha = 0.05$.)

8. A random sample of 100 students was taken to test the claim that the mean amount of time college students spend watching television per week is greater than 14.5 hours. The resulting values yielded a mean of 16.3 hours. Complete the test using $\sigma = 5.2$ hours and $\alpha = 0.02$.

9. Estimate the mean monthly rental fee for second-floor one-bedroom apartments in our city, with a 90 percent confidence interval. A random sample of 12 such rental fees resulted in a sample mean of $138.50. (Use σ = $9.00.)

10. Find a 95 percent confidence-interval estimation of the mean length of time required to complete an examination based on a sample of 50 students who required a mean time of 115 minutes. (Use σ = 8 minutes.)

INFERENCES INVOLVING ONE POPULATION

9

LESSON 9-1: INFERENCES CONCERNING ONE MEAN

In Chapter 8 you were introduced to a five-step procedure called a hypothesis test and to the idea of a confidence-interval estimate. Both of these inferences required applying the concept of a sampling distribution, as will all the other inferences to be discussed in Chapters 9 and 10. Although each of the situations will require its own sampling distribution and the use of a corresponding set of formulas, each hypothesis test will follow the same five steps, with an appropriate adjustment made in the formulas. Each confidence-interval estimate will follow the same basic pattern of construction, except for the specific formula that is to be used.

In Chapter 8 we applied the central limit theorem (CLT) to the sampling distribution of $\bar{x}$'s. Recall that the CLT stated that the distribution of sample means, $\bar{x}$'s, is normally distributed about the mean of the population, μ, with a standard deviation of $\sigma/\sqrt{n}$ ("standard error of the mean").

Note: The standard error of any sample statistic is the standard deviation of the sampling distribution for that sample statistic. Standard error is used to identify it as being different from the standard deviation of individuals from either the population or the sample.

The test statistic z was calculated in step 4 of the procedure outlined in Chapter 8. The observed value of the standard score, z^*, is then compared to the critical value(s) as identified in step 3, in order to reach a decision. In a hypothesis-testing situation, where one mean is being tested, the standard deviation is seldom known: thus the formula $z = (\bar{x} - \mu)/(\sigma/\sqrt{n})$ cannot be used since the standard error of the mean, $\sigma/\sqrt{n}$, cannot be accurately determined. The best possible replacement that can be found is $s/\sqrt{n}$, since s, the standard deviation of the data in the observed sample, is the only estimate available for σ, the population standard deviation. When this approximation is used, we must make some other adjustments also. One of these is the use of the t-test statistic. t is a standard score, much like z, except that it is the test statistic used whenever the standard error of an otherwise normal sampling distribution is to be approximated. In general, t will be defined by:

128 INFERENCES: ONE POPULATION   Chapter 9

$$t = \frac{\text{statistic} - \text{parameter}}{\text{standard error of statistic}}$$

(The statistic is the sample statistic, like $\bar{x}$. The parameter is the corresponding population parameter, and μ corresponds to $\bar{x}$. The standard error is the standard deviation of the sampling distribution of repeated values of the sample statistic.) Thus whenever $s/\sqrt{n}$ is used as the estimate for $\sigma/\sqrt{n}$, then the test statistic is t.

$$t = \frac{\bar{x} - \mu}{s/\sqrt{n}} \qquad \text{[formula (9-1), p. 328 in textbook]}$$

This t statistic is approximately normally distributed and belongs to a distribution known as "Student's t distribution." In using the t-test statistic, we will follow exactly the same procedure as when using the z score. The distinguishing feature is whether σ is known or unknown.

Note: Use z whenever σ is known, and use t whenever σ is unknown.

The use of t, as indicated above, results in the use of an approximation. Besides making this approximation, we must contend with making a judgment as to the "goodness" of this approximation. A measure of "worth" was introduced by Student (the pen name for W. S. Gosset) and is called the "number of degrees of freedom" (df). The better the worth of the estimate, the higher the number of degrees of freedom. For inferences about one mean, this worth is directly related to the size of the sample being used. df is defined to be $n - 1$.

Inspection of the table of critical values for Student's t distribution will suggest that a different distribution occurs for each different number of degrees of freedom possible, and this is true. If a sample statistic like s has a "worth" of df = 5, then the critical value of t is set appropriately. Notice also that the critical values of t approach the critical values of z as the number of degrees of freedom increases. It is typical for statisticians to consider all cases of df greater than or equal to 30 as one distribution. Once df is 30 or larger, the t distribution is barely distinguishable from the normal distribution (z). Thus for all situations where df is greater than 29, we will actually use the same critical values as we use for z. This only means that the critical values happen to be the same — t and z are still different test statistics, t being used when s is used to approximate σ, and so on, as defined previously.

All confidence-interval estimate formulas that use the CLT applications have the form:

statistic ± (critical value of test statistic) · (standard error of statistic)

and the confidence interval becomes:

L.C.L. < parameter < U.C.L.

where the parameter is the population parameter being estimated, the statistic is the corresponding sample statistic (the point estimate for the parameter), while

ONE PROPORTION Lesson 9-2

the critical values for the test statistic are the values obtained from the tables, and the standard error is the standard deviation of the sampling distribution of the sample statistic. L.C.L. and U.C.L. are the lower and upper confidence-interval limits found by performing the subtraction and addition, respectively, in the formula. Again, z is used when the value of the standard error is known, while t will be used whenever the standard error is estimated. Number of degrees of freedom, critical values, and so on, are determined as they were for the hypothesis test.

Specifically, the formulas for the two inferences about one mean value are:

$$t = \frac{\bar{x} - \mu}{s/\sqrt{n}}$$ [formula (9-1), p. 328 in textbook]

and

$$\bar{x} \pm t(df, \alpha/2) \cdot \frac{s}{\sqrt{n}}$$ [formula (9-2), p. 333 in textbook]

EXERCISES 9-1

1. Describe how we distinguish between situations where t and z are used as the test statistic.

2. What is the "standard error of the mean"? Describe it and give two formulas that can be used for calculating its value. Why are there two formulas?

3. What general form do both the t and z test statistics fit?

4. What general form does a confidence-interval formula using t or z fit?

LESSON 9-2: INFERENCES CONCERNING ONE PROPORTION

Inferences concerning one proportion of a population follow the same basic pattern as was discussed in Lesson 9-1. The difference is in the sample statistic and its corresponding formulas. In dealing with inferences about proportions, we actually have two different sample statistics. The binomial random variable x, which is a count of the number of times a particular property occurred in a sample of n objects, is one sample statistic. This binomial random variable is easily converted to the second sample statistic, the observed binomial probability p' ($p' = x/n$). We use the distribution of the repeated values of this sample statistic, p', as our sampling distribution. The mean value of p' is p, the true value of the probability of occurrence within the population. (Thus p' is a sample statistic and p is its corresponding population parameter.) The standard deviation of p' is given by the formula:

130 INFERENCES: ONE POPULATION   Chapter 9

$$\sigma_{p'} = \sqrt{\frac{pq}{n}} \qquad \text{(standard error of proportion)}$$

as described in the textbook on page 337. The distribution of p' is approximately normal. On page 337 in the textbook there are guidelines that describe the conditions under which the binomial variable x is approximately normal. p' is only a multiple (the $1/n$ multiple of x; therefore, it will be approximately normal under the same conditions. This approximation improves as the sample size increases. Since it is typical for these sample sizes to be quite large, we will consider p' to be normally distributed.

Our formulas for inference making, then, are:

$$z = \frac{p' - p}{\sqrt{pq/n}} \qquad \text{[formula (9-3), p. 337 in textbook]}$$

and

$$p' \pm z(\alpha/2)\sqrt{\frac{p'q'}{n}} \qquad \text{[formula (9-4), p. 339 in textbook]}$$

where $p' = x/n$ and $q' = 1 - p'$. Formula (9-3) is for the hypothesis test, and p (both in the numerator and denominator) is the value claimed by the null hypothesis being tested. (Remember, H_0 is assumed to be true until evidence indicates otherwise.) Notice that in formula (9-4) you do not find p; instead, you find p' used for the calculation of the standard error. p is unknown and that is why we are estimating its value. If it were known, there would be no need for the confidence interval of estimation.

We commented above that the sample size for a problem of this nature was usually quite large. There are two basic reasons for this. (1) The data are typically easy to obtain. (Usually you would just inspect the subject for a certain characteristic — does the subject display this characteristic, yes or no?) (2) Unless the sample is fairly large the information found is virtually worthless. To demonstrate this second point, let's look at the following illustration.

Illustration   Construct the 95 percent confidence interval of estimation for the true proportion of heads when the coin tossed showed 4 heads in 10 tosses.

Solution

$$p' \pm z(\alpha/2) \cdot \sqrt{\frac{p'q'}{n}}, \; p' = \frac{x}{n} = \frac{4}{10} = 0.4, \; q = 1 - 0.4 = 0.6,$$

$$z(\alpha/2) = z(0.025) = 1.96 \approx 2.0.$$

$$0.4 \pm (2.00)\sqrt{\frac{(.4)(.6)}{10}}$$

$$0.4 \pm (2.00)\sqrt{0.0240}$$

$$0.4 \pm (2.0)(0.15)$$

ONE STANDARD DEVIATION   Lesson 9-3

$$0.4 \pm 0.3$$

$$0.1 < p < 0.7$$

Stop and think — how much was learned by constructing the interval estimate based on a sample of size 10? p has to be between zero and one by definition. To say that it is somewhere between 0.1 and 0.7 does not narrow it down very much. These results hardly seem worth the effort of a sample; you could probably guess as close. Therefore, if the estimate is to be reasonably close to the true value, then we must take a larger sample.

Notice that the z statistic is used here even when the standard error, $\sigma_{p'}$, is being estimated. This is because of the closeness of approximation to the normal. Whenever n is greater than 20 and np and nq are both larger than 5, the distribution is close enough to the normal that there is no need to go to Student's t distribution. And as indicated above, typically n will be much larger than 20.

EXERCISES 9-2

1. What is the standard error of proportion? Describe it and give two of the formulas used for its calculation. Why are there two formulas? When is each used?

2. What test statistic is used for the inferences involving proportion?

LESSON 9-3: INFERENCES CONCERNING ONE STANDARD DEVIATION

When dealing with the sample statistic standard deviation s, we do not use a sampling distribution in the same manner as we did with means and proportions. Neither the sample variance nor the sample standard deviation have distributions that are normal. As it turns out, we convert either sample statistic into another quantity. Let's call it Q for the moment and define Q to be the number

$$\frac{(n-1)s^2}{\sigma^2}$$

where n is the size of the sample being used, s^2 is the variance observed within that sample, and σ^2 is the variance of the population from which the sample was drawn. As a result of this "coding" of the sample statistic, we can now describe a "sampling distribution" of Q's. The resulting distribution of Q's is a chi-square (χ^2) distribution.

The chi-square distribution is a theoretical probability distribution, as is the normal distribution. The distribution of chi square is nonnegative and is skewed to the right. Its mean is always the same as the numerical value of the numbers of degrees of freedom associated with each distribution. These properties of the chi-square distribution are discussed in the textbook on pages 346 and 347,

INFERENCES: ONE POPULATION   Chapter 9

where the distribution is described by inspecting the value of Q, mentioned above. Since Q has this χ^2 distribution, we replace Q by χ^{2*} (* indicates the calculated value of the test statistic) and we will calculate the observed value of the chi square by using formula (9-7).

$$\chi^{2*} = \frac{(n-1)s^2}{\sigma^2}$$ [formula (9-7), p. 346 in textbook]

(There are other statistical quantities that have a chi-square distribution, as we will see in Chapter 11.)

In hypothesis tests for standard deviation or variance, we will calculate χ^{2*} and compare it to the critical value(s) obtained from Table 7. Otherwise the test procedure follows the same basic five steps.

To construct the $1 - \alpha$ percent confidence-interval estimate for either σ or σ^2, we would first solve formula (9-7) for σ^2:

$$\sigma^2 = \frac{(n-1)s^2}{\chi^2}$$

In place of χ^2 we will substitute the two corresponding critical values of chi square from Table 7, $\chi^2(df,\alpha/2)$ and $\chi^2(df,1-\alpha/2)$. The two values obtained will be $A = (n-1)s^2/\chi^2(df,\alpha/2)$ and $B = (n-1)s^2/\chi^2(df,1-\alpha/2)$.

Since $\chi^2(df,\alpha/2)$ is larger than $\chi^2(df,1-\alpha/2)$, as seen in Figure 9-1, the

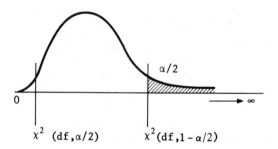

Figure 9-1

value of A will be smaller than the value of B. (Equals divided by unequals are unequal in the opposite order.) Thus the confidence-interval formula for the estimation of σ^2 is:

$$\frac{(n-1)s^2}{\chi^2(df,\alpha/2)} \text{ to } \frac{(n-1)s^2}{\chi^2(df,1-\alpha/2)}$$ [formula (9-9), p. 351 in textbook]

Notice that this formula does not have the form of the two previous confidence-interval formulas (χ^2 is not a symmetrical distribution as are t and z).

The confidence-interval estimate for the population standard deviation is found by taking the square root of each of the two terms of formula (9-9).

The result is formula (9-10).

$$\sqrt{\frac{(n-1)s^2}{\chi^2(df,\alpha/2)}} \text{ to } \sqrt{\frac{(n-1)s^2}{\chi^2(df,1-\alpha/2)}}$$

EXERCISES 9-3

1. What test statistic is used for making inferences about the variance or standard deviation of one population? Describe its pattern of behavior.

SELF-CORRECTING EXERCISES FOR CHAPTER 9

1. For each item, state the null hypothesis (H_0) and the alternative hypothesis (H_a).
   a. The standard deviation of the lengths of fish "kept" by fishermen who fished Cayuga Lake last year was no more than 4.5 inches.
   b. The proportion of students at our college who wear glasses is at least 40 percent.
   c. Seven percent of the students at our college obtain medical exemption from some part of their physical education requirement.
   d. The variance of the heights of students enrolled in liberal arts at our college is less than 15.
   e. At least 20 percent of the cars driven on our streets are foreign-made.
   f. The mean length of infants born at General Hospital is 19.2 inches.
   g. The variance of the grades received on a certain exam was more than 150.

2. Using Table 6 on page 574 in the textbook, determine the critical values for Student's t (names and values) shown in the following sketches.

a.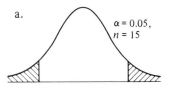
$\alpha = 0.05$, $n = 15$

b.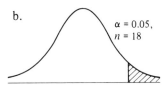
$\alpha = 0.05$, $n = 18$

c.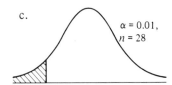
$\alpha = 0.01$, $n = 28$

d.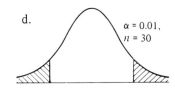
$\alpha = 0.01$, $n = 30$

e.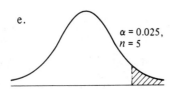
$\alpha = 0.025$, $n = 5$

f.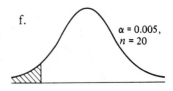
$\alpha = 0.005$, $n = 20$

g.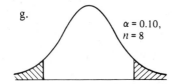
$\alpha = 0.10$, $n = 8$

h.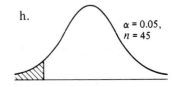
$\alpha = 0.05$, $n = 45$

3. Using Table 6 on page 574 in the textbook, determine each of the following critical values of the Student's t distribution.
   a. $t(20, 0.05)$ b. $t(5, 0.01)$ c. $t(14, 0.005)$
   d. $t(12, 0.025)$ e. $t(50, 0.05)$ f. $t(15, 0.95)$
   g. $t(20, 0.95)$ h. $t(50, 0.99)$

4. Determine the test criteria (test statistic, critical value(s), and critical region(s)) that would be used in testing each of the following hypothesis tests.
   a. H_0: $\mu = 15.0$ (σ unknown, $n = 15$, $\alpha = 0.05$)
      H_a: $\mu \neq 15.0$

   b. H_0: $\mu = 12.6$ (σ unknown, $n = 18$, $\alpha = 0.01$)
      H_a: $\mu < 12.6$

   c. H_0: $\mu = 1500$ (σ unknown, $n = 10$, $\alpha = 0.02$)
      H_a: $\mu \neq 1500$

   d. H_0: $\mu = 56.2$ (σ unknown, $n = 50$, $\alpha = 0.05$)
      H_a: $\mu > 56.2$

   e. H_0: $\mu = 3.05$ (σ unknown, $n = 20$, $\alpha = 0.01$)
      H_a: $\mu > 3.05$

   f. H_0: $\mu = 76.1$ (σ unknown, $n = 25$, $\alpha = 0.05$)
      H_a: $\mu < 76.1$

5. Using Table 7 on page 575 in the textbook, determine the critical values for the chi-square distribution (names and values) shown in the following sketches.

SELF-CORRECTING EXERCISES

a. 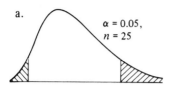 $\alpha = 0.05$, $n = 25$

b. 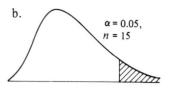 $\alpha = 0.05$, $n = 15$

c. 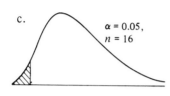 $\alpha = 0.05$, $n = 16$

d. 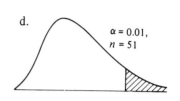 $\alpha = 0.01$, $n = 51$

e. 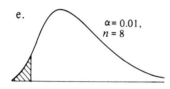 $\alpha = 0.01$, $n = 8$

f. 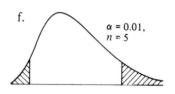 $\alpha = 0.01$, $n = 5$

g. 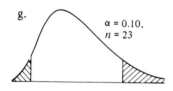 $\alpha = 0.10$, $n = 23$

h. 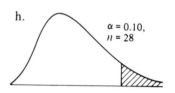 $\alpha = 0.10$, $n = 28$

6. Using Table 7 on page 575 in the textbook, determine each of the following critical values of the chi-square distribution.
   a. $\chi^2(25, 0.05)$
   b. $\chi^2(28, 0.99)$
   c. $\chi^2(5, 0.10)$
   d. $\chi^2(17, 0.025)$
   e. $\chi^2(21, 0.90)$
   f. $\chi^2(2, 0.05)$
   g. $\chi^2(15, 0.95)$
   h. $\chi^2(18, 0.975)$

7. Using Table 5 on page 573 of the textbook, determine the following critical values for the standard normal score z.

a. 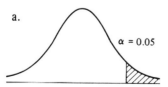 $\alpha = 0.05$

b. 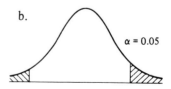 $\alpha = 0.05$

c. 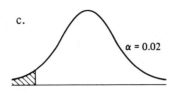   d.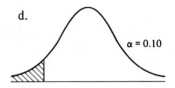

e. $z(0.05)$ f. $z(0.02)$
g. $z(0.99)$ h. $z(0.975)$

8. Determine the text criteria (test statistic, critical value(s), and critical region) that would be used in a hypothesis test of each of the following cases.

a. H_0: $p = 0.45$ ($\alpha = 0.05$)
   H_a: $p \neq 0.45$

b. H_0: $p = 0.98$ ($\alpha = 0.02$)
   H_a: $p > 0.98$

c. H_0: $p = 0.05$ ($\alpha = 0.10$)
   H_a: $p < 0.05$

d. H_0: $\sigma = 0.5$ ($n = 15, \alpha = 0.05$)
   H_a: $\sigma < 0.5$

e. H_0: $\sigma^2 = 10.5$ ($n = 25, \alpha = 0.02$)
   H_a: $\sigma^2 \neq 10.5$

f. H_0: $\sigma = 105.2$ ($n = 75, \alpha = 0.10$)
   H_a: $\sigma > 105.2$

9. a. If the value of the test statistic falls in the critical region, what decision is reached?
   b. If the value of the test statistic does not fall in the critical region, what decision is reached?

10. "The average American smoker consumes at least 4,100 cigarettes per year." Test this claim using the following random sample information.

$n = 400$ $\bar{x} = 4,015$ $s = 1,125$ $\alpha = 0.05$

11. Estimate the cost of dinner for two when out on the town in a major city, with a 90 percent confidence interval, using the sample listed below.

Evening Dining Costs in Major Cities

City	Est. Cost for 2
Atlanta	$49.00
Boston	44.00
Chicago	48.50
Dallas	47.00
Detroit	49.00
Los Angeles	47.00
New York	59.00
Philadelphia	47.50
San Francisco	52.50
Washington, D.C.	49.00

Definition — Cost is for dinner (prime ribs or house specialty) for two persons at a white-tablecloth restaurant, including two drinks apiece, one bottle wine, one cordial each, tax, and 15% gratuity.

12. A random sample of 400 married women were asked, "Did your honeymoon live up to all your expectations?" Three hundred and seventy-two responded NO. Use this data to test the hypothesis that 95 percent of all honeymoons do not measure up to the brides' expectations. (Use $\alpha = 0.01$.)

13. In a national sample of 500 school-age boys, 61 boys revealed that they had at least once in the year prior to the interview committed the offense of carrying a concealed weapon. Using this sample data, construct the 95 percent confidence-interval estimate for the true proportion of all school-age boys who have committed this offense in the last year.

14. Test the claim that the standard deviation of scores on last semester's mathematics final exam was greater than 12. The scores obtained by 100 randomly selected students had a standard deviation of 13.75. (Use $\alpha = 0.05$.)

15. Construct the 90 percent confidence interval for the estimation of the variance among the weights of college women based on a random sample of 20 college women whose weights had a mean of 119.5 pounds and a standard deviation of 11.9 pounds.

INFERENCES INVOLVING TWO POPULATIONS

10

LESSON 10-1: INFERENCES CONCERNING THE DIFFERENCE BETWEEN TWO INDEPENDENT MEANS

Inferences concerning the difference between the means of two independent normally distributed populations are made by using the central limit theorem. The CLT applied here tells us that the sampling distribution of the differences between two sample means ($\bar{x}_1 - \bar{x}_2$) is normally distributed about a mean equal to the difference between the population means ($\mu_1 - \mu_2$) with a standard deviation of $\sqrt{\sigma_1^2/n_1 + \sigma_2^2/n_2}$ (the standard error of difference between independent means). Therefore, when σ_1 and σ_2 are known, the formula to be used for the hypothesis test is:

$$z = \frac{(\bar{x}_1 - \bar{x}_2) - (\mu_1 - \mu_2)}{\sqrt{(\sigma_1^2/n_1) + (\sigma_2^2/n_2)}} \qquad \text{[formula (10-1), p. 367 in textbook]}$$

The formula for the confidence estimation of the difference $\mu_1 - \mu_2$ will be found by:

$$(\bar{x}_1 - \bar{x}_2) \pm z(\alpha/2) \sqrt{\frac{\sigma_1^2}{n_1} + \frac{\sigma_2^2}{n_2}} \qquad \text{[formula (10-2), p. 369 in textbook]}$$

In this situation the standard error will be known only when the values of both population standard deviations are known; then we will use the standard normal score z for the test statistic. In this lesson, this is the only case of independent means that is considered, but in lesson 10-3 we will look at the other cases (σ's not both known) and learn to apply the z or t statistic as required.

Let's look at the formula for the standard error $\sigma_{\bar{x}_1 - \bar{x}_2}$ again. Does it seem reasonable that its value should be found by the following formula?

$$\sigma_{\bar{x}_1 - \bar{x}_2} = \sqrt{\frac{\sigma_1^2}{n_1} + \frac{\sigma_2^2}{n_2}}$$

Does the addition seem reasonable? Let us suggest a situation and use a measure of variation that should help us answer this question. Consider some kind of assembly where a machined shaft has a bearing and shims placed on it in a specified interval. The shaft has 0.75 ± 0.02 inches reserved for the bearing and the shims. Let x represent the width of this space ($0.73 < x < 0.77$). The bearing that is to occupy this space has a width, y, of 0.70 ± 0.01 inches ($0.69 < y < 0.71$). The shims are to take up whatever space is left, $x - y$. Let's consider the range as our measure of variation. The range of the x measurements is 0.04 ($0.77 - 0.73 = 0.04$), while the range of the y measurements is 0.02 ($0.71 - 0.69 = 0.02$). The selection of what bearings go on which shaft is strictly random. Now let's consider the shim. $x - y$ is the amount of space to be shimmed. The amount of shim material needed will range from 0.02 to 0.08 inches. A range of 0.06 inches is needed for the variable $x - y$. (Least amount of shim will be used when the narrowest x and the widest y occur together. The most shim will be used when widest x and narrowest y occur together.) Notice that the range of $x - y$ is the sum of the ranges for x and y.

The range is not quite the same as the variance; however, both do measure the amount of spread or dispersion found in the data. This simple illustration should at least convince you that the sum of the two respective measures is reasonable. The "sum" because the formula $\sqrt{\sigma_1^2/n_1 + \sigma_2^2/n_2}$ could be rewritten as $\sqrt{(\sigma_1/\sqrt{n_1})^2 + (\sigma_2/\sqrt{n_2})^2}$. (The variances are added.) This would then seem quite comparable to $\sigma/\sqrt{n}$, which is the standard error of the mean studied earlier.

EXERCISES 10-1

1. Why is the z statistic used for the case of inferences mentioned above about the difference between two independent means?

2. Do formulas (10-1) and (10-2) fit the general forms discussed in Exercises 3 and 4 of Lesson 9-1?

LESSON 10-2: INFERENCES CONCERNING TWO VARIANCES

When making inferences concerning the variances (or standard deviations) of two independent normal populations, we will use the ratio of variances (not difference, as is used with means) and a test statistic that is used only for this kind of inference. As we saw in Chapter 9 and in Lesson 9-3, the variance or standard deviation of a sample behaves differently from other sample statistics. The ratio of two observed sample variances is the defining equation for the F distribution: $F = s_1^2/s_2^2$, where s_1^2 and s_2^2 are the variances of two independent samples of specified size drawn from two populations of equal variance. Since s^2 by definition is nonnegative, the F distribution will be nonnegative and looks like Figure 10-1.

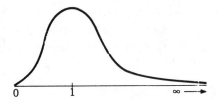

Figure 10-1

This sampling distribution is used to test the null hypothesis $\sigma_1^2/\sigma_2^2 = 1$, and to construct confidence-interval estimates on the ratio of two sample variances or standard deviations. Each pair of sample sizes generates a different sampling distribution for F. Thus the critical values of F are determined by three identifying values, df_n, df_d, and α, where df_n and df_d are the number of degrees of freedom for the sample whose variance is used for the numerator and denominator, respectively, and α is the area under the curve to the right of the point in question. Consequently, several special adaptations are used in order to tabulate the critical values as conveniently as possible. The details for these adaptations may be found in the textbook on pages 374-380.

EXERCISES 10-2

1. In what way are two variances or standard deviations compared?

2. What test statistic is used for comparing two variances?

3. How do you determine the critical value from the tables for the F distribution?

LESSON 10-3: INFERENCES CONCERNING THE DIFFERENCE BETWEEN TWO INDEPENDENT MEANS

In Chapter 10 and in Lesson 10-1 you were introduced to the basic concepts concerning the inferences about the difference between two independent means. The information presented there revolved around the fact that the values of both population standard deviations were known or both sample sizes are large. We are now ready to investigate what happens when the standard error, $\sigma_{\bar{x}_1 - \bar{x}_2}$, needs to be estimated. The first thing that you should recognize is that the t-test statistic will be used and will be defined by:

$$t = \frac{(\bar{x}_1 - \bar{x}_2) - (\mu_1 - \mu_2)}{\text{estimate of } \sigma_{\bar{x}_1 - \bar{x}_2}}$$

We want to use the best possible estimate for the standard error so that our results are reliable.

In order to achieve the best possible estimate for the standard error, we will have to determine exactly which one of three possible cases we are dealing with.

DIFFERENCE BETWEEN TWO INDEPENDENT MEANS   Lesson 10-3

They all have the property that one or both values of σ (σ_1 and σ_2) are not known. From there they may be classified as: (1) both samples are large, (2) one or both samples are small and have equal variances, or (3) one or both samples are small and have unequal variances.

The first step in determining which case we have is very easy. Simply observe the sample sizes. If both are larger than or equal to 30 we have case 1. If both are not larger than 30, then we have either case 2 or 3. To decide between cases 2 and 3, we will use an F test for the equality of the two respective population variances. A decision of "fail to reject H_0" will lead us to case 2, while rejecting H_0 will give us case 3. Figure 10-2 shows a tree diagram depicting this sequence of decisions.

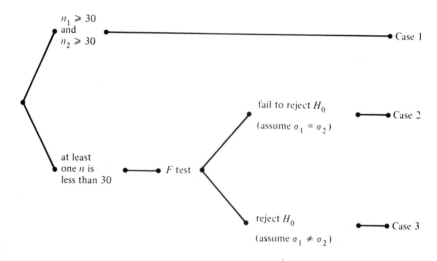

Figure 10-2

The formula for the estimate of the standard error is given below for each of the three cases.

	Standard error	Special formulas
Case 1:	$\sqrt{\dfrac{s_1^2}{n_1} + \dfrac{s_2^2}{n_2}}$	Use z as shown in Lesson 10-1 and Section 10-2 of textbook.
Case 2:	$s_p \sqrt{\dfrac{1}{n_1} + \dfrac{1}{n_2}}$	$s_p = \sqrt{\dfrac{(n_1 - 1)s_1^2 + (n_2 - 1)s_2^2}{n_1 + n_2 - 2}}$ and df $= n_1 + n_2 - 2$
Case 3:	$\sqrt{\dfrac{s_1^2}{n_1} + \dfrac{s_2^2}{n_2}}$	df = the smaller of $n_1 - 1$ or $n_2 - 1$

INFERENCES: TWO POPULATIONS Chapter 10

In case 1, where both samples are large, the estimate for the standard error is sufficiently accurate that the sampling distribution is best estimated by use of the normal distribution. Therefore, we will use the z test statistic.

In case 2, s_p represents the pooled standard deviation for the two samples. "Pooled" essentially means that the two samples were put together as one sample in order to get the best possible estimate for σ. (Remember, this case assumed that $\sigma_1 = \sigma_2$.) From this point on, the testing of a hypothesis or an estimation follows our familiar routine.

EXERCISES 10-3

1. Describe how we distinguish between the four classifications of the difference between two independent means.

LESSON 10-4: INFERENCES CONCERNING THE DIFFERENCE BETWEEN TWO DEPENDENT MEANS

The difference between the mean values of two dependent populations is treated as the mean value of the difference in values of the paired data. That is, instead of making inferences about the difference between the means ($\mu_x - \mu_y$), we will make inferences about the mean value of paired differences, μ_d, where $d = x - y$. Recall that the "dependent mean" is the mean of the differences between two populations of values stemming from common sources. Thus each source element in the population of objects is supplying two response values, x and y. Each of these pairs of values will be used to generate a single variable d, the paired difference. This allows us to compare each x-value with its corresponding y-value instead of directly comparing the mean value of x with the mean value of y. As described in the textbook, each of these two techniques has its own set of advantages and disadvantages. The situation at hand will dictate the method to be used.

The use of the dependent samples (or paired difference) generates a situation where we actually have only one variable to work with. That one variable is d. We will use our sample of n values of d to obtain an observed mean value, $\bar{d}$, and an observed standard deviation for the d's, s_d. The central limit theorem can be applied to the sampling distribution of d's. The $\bar{d}$'s will be normally distributed about the true mean value of d, μ_d, with a standard deviation of $\sigma_d/\sqrt{n}$.

The standard error of $\bar{d}$ will have to be estimated, and therefore is estimated by $s_d/\sqrt{n}$. With this estimation, we use the t statistic and these familiar formulas:

$$t = \frac{\bar{d} - \mu_d}{s_d/\sqrt{n}} \qquad \text{[formula (10-17), p. 393 in textbook]}$$

DIFFERENCE BETWEEN TWO PROPORTIONS Lesson 10-5

and

$$\bar{d} \pm t(\text{df}, \alpha/2)\,(s_d/\sqrt{n}) \qquad \text{[formula (10-18), p. 394 in textbook]}$$

The completion of the inferences about the difference between dependent means is very comparable to the inferences about one mean value when σ is unknown. You should be able to see why this is the case. There are no formulas given for the use of z here, since the standard error for the difference between two dependent means, σ_d, will almost never be known.

EXERCISES 10-4

1. What test statistic is used for inferences about the difference between dependent means? Explain why.
2. Do formulas (10-17) and (10-18) fit the general forms for formulas using t as discussed in Exercises 3 and 4 in Lesson 9-1?

LESSON 10-5: INFERENCES CONCERNING THE DIFFERENCE BETWEEN TWO PROPORTIONS

The z statistic is used with inferences concerning the difference between two proportions, since the sampling distribution of $p_1' - p_2'$ is approximately normal. (Proportions are binomial probabilities.) The mean of the sampling distribution is $p_1 - p_2$, the difference between the true values of the respective proportions. The standard error is obtained by the formula:

$$\sqrt{p^* q^* \left(\frac{1}{n_1} + \frac{1}{n_2}\right)}$$

where p^* is the best possible estimate we have for the pooled proportion. This value for p^* can come from one of two sources for a hypothesis test: (1) $p^* = p_1 = p_2$, or (2) $p^* = (x_1 + x_2)/(n_1 + n_2)$. The first case will be used when the two populations involved have stated their value for p. For example, population 1 claims $p = 0.75$ and population 2 claims its proportion is the same, 0.75. The second formula would be used when the two populations state how their proportions are equal, but do not specify the value involved. For example, manufacturer 1 says that the proportion of his product that surpasses its guarantee period without malfunction is the same as or higher than manufacturer 2's proportion. Notice, in this last example there is no claim as to the value of p_1 or p_2; therefore, the sample data will have to be used to estimate this proportion. Notice that formula 2, above, takes the total of both samples (pools them) in order to obtain the best possible single estimate. The confidence-interval estimate will always use the second formula, the sample data, for its estimate.

144 INFERENCES: TWO POPULATIONS   Chapter 10

EXERCISES 10-5

1. Show that the formula for $\sigma_{p'_1 - p'_2}$ is consistent with the remarks made in Lesson 10-1 about the addition of variances of the separate distributions to get the standard error of the sampling distribution of differences.

SELF-CORRECTING EXERCISES FOR CHAPTER 10

1. For each item state the null hypothesis (H_0) and the alternative hypothesis (H_a).
   a. There is no difference between the proportion of male students and the proportion of female students who obtain a medical exemption from the physical education requirement at our college.
   b. The variance in the length of infant girls and the variance in the length of infant boys are the same.
   c. The standard deviation of the weights of males is greater than the standard deviation of the weights of females in the adult population of our county.
   d. The proportion of blue cars seen on the street is at least 10 percent higher than the proportion of red cars.
   e. There is no difference between the mean grade-point average of this year's graduating class and the mean grade-point average of last year's class.
   f. The mean of the difference in the agility scores attained by police science majors before and after taking their physical fitness course is positive.
   g. The average American-made car weighs 1,000 pounds more than the average foreign-made car.

2. Using Table 8 on pages 576–581 in the textbook, determine the following critical values for the F distribution.

a. 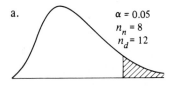   b.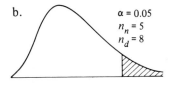

c. 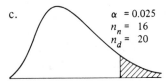   d.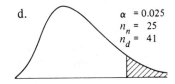

SELF-CORRECTING EXERCISES

e. 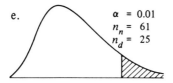   f.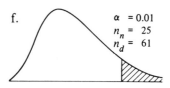

g. 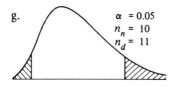   h.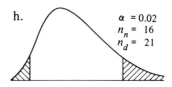

3. Using Table 8 on pages 576–581 of the textbook, determine each of the following critical values for the F distribution.
   a. $F(5, 8, 0.05)$
   b. $F(15, 30, 0.025)$
   c. $F(60, 40, 0.01)$
   d. $F(12, 35, 0.05)$
   e. $F(18, 25, 0.05)$
   f. $F(35, 45, 0.05)$
   g. $F(6, 10, 0.95)$
   h. $F(12, 20, 0.99)$

4. Determine the names and the values of each of the following critical values.

a. 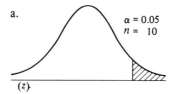   b.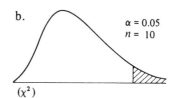

c. 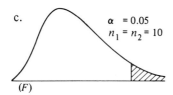   d.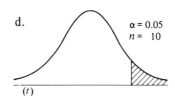

e.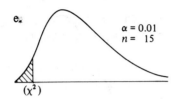
$\alpha = 0.01$
$n = 15$
(χ^2)

f.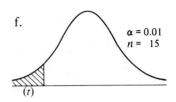
$\alpha = 0.01$
$n = 15$
(t)

g.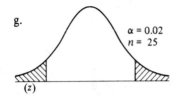
$\alpha = 0.02$
$n = 25$
(z)

h.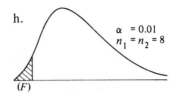
$\alpha = 0.01$
$n_1 = n_2 = 8$
(F)

i.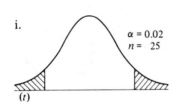
$\alpha = 0.02$
$n = 25$
(t)

j.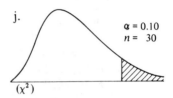
$\alpha = 0.10$
$n = 30$
(χ^2)

5. Determine the test criteria (test statistic, critical value(s), and critical region) that would be used in a hypothesis test of each of the following cases.

   a. $H_0: \mu_A - \mu_B = 0$ (σ known, $\alpha = 0.05$)
   $H_a: \mu_A - \mu_B \neq 0$

   b. $H_0: \mu_I - \mu_{II} = 0 \; (\leqslant)$ (σ known, $\alpha = 0.10$)
   $H_a: \mu_I - \mu_{II} > 0$

   c. $H_0: \mu_c - \mu_p = 50 \; (\geqslant)$ (σ known, $\alpha = 0.01$)
   $H_a: \mu_c - \mu_p < 50$

   d. $H_0: \sigma_A^2 = \sigma_B^2 \; (\leqslant)$ ($n_A = 9, n_B = 12, \alpha = 0.05$)
   $H_a: \sigma_A^2 > \sigma_B^2$

   e. $H_0: \sigma_I^2 = \sigma_{II}^2$ ($n_A = 10, n_B = 10, \alpha = 0.05$)
   $H_a: \sigma_I^2 \neq \sigma_{II}^2$

   f. $H_0: \sigma_A^2 / \sigma_B^2 = 1 \; (\leqslant)$ ($n_A = n_B = 25, \alpha = 0.01$)
   $H_a: \sigma_A^2 / \sigma_B^2 > 1$

g. $H_0: \mu_A - \mu_B = 0$ (σ's unknown, but assume $\sigma_A = \sigma_B$,
   $H_a: \mu_A - \mu_B \neq 0$ $n_A = 45, n_B = 55, \alpha = 0.10$)

h. $H_0: \mu_A = \mu_B (\leq)$ (σ's unknown, but assume $\sigma_A = \sigma_B$,
   $H_a: \mu_A > \mu_B$ $n_A = 10, n_B = 12, \alpha = 0.05$)

i. $H_0: \mu_I - \mu_{II} = 10 (\geq)$ (σ's unknown but equal, $n_I = 25$,
   $H_a: \mu_I - \mu_{II} < 10$ $n_{II} = 36, \alpha = 0.02$)

j. $H_0: \mu_d = 0$ ($n = 12, \alpha = 0.01$)
   $H_a: \mu_d \neq 0$

k. $H_0: \mu_d = 5 (\leq)$ ($n = 24, \alpha = 0.05$)
   $H_a: \mu_d > 5$

l. $H_0: \mu_d = 0 (\geq)$ ($n = 50, \alpha = 0.10$)
   $H_a: \mu_d < 0$

m. $H_0: p_L - p_M = 0 (\leq)$ ($\alpha = 0.05$)
   $H_a: p_L - p_M > 0$

n. $H_0: p_F - p_M = 0.15 (\leq)$ ($\alpha = 0.10$)
   $H_a: p_F - p_M > 0.15$

6. Sunset and Happy Days bowling leagues are comparable leagues in two different cities. Test the hypothesis that there is no real difference between these two leagues, using the results shown below. (Use $\sigma = 4.3$ and $\alpha = 0.04$.)

	n	$\bar{x}$
Sunset	20	158.3
Happy Days	25	160.7

7. The difference between the mean life of light bulbs produced by two different manufacturers is to be estimated by a 95 percent confidence interval. Two random samples are obtained and tested. Construct the confidence interval.

Manufacturer A: $\sigma = 120$ hours $\bar{x} = 1450$ $n = 150$
Manufacturer B: $\sigma = 90$ hours $\bar{x} = 1275$ $n = 100$

8. "The good students get better while the poor students get poorer." A class of 25 elementary school children was given a reading-achievement test recently. The mean and the standard deviation of the raw scores were 50.3 and 15.6. Two years ago they obtained a mean of 38.7 and a standard deviation of 10.6

on a similar exam. During the two-year period the mean has increased; but notice that the standard deviation has also increased. Using $\alpha = 0.05$, test the claim that the standard deviation is now significantly greater than before.

9. "Top-Flite proves again; it's the longest ball you can play." Thirty-eight professional golfers hit hundreds of wood and iron shots for distance. The set of data shows Top-Flite against a composite of 10 other leading brands of golf balls.

	Number of drives	Average distance	Standard deviation
Top-Flite	76	215.5	15.6
Others	190	205.5	17.6

Test the claim stated above using mean driving distance as the measure of long ball. (Use $\alpha = 0.05$.)

10. Estimate the difference between the mean bowling averages of bowlers in class A and class B divisions of a men's industrial league using the following sample results. ($1 - \alpha = 0.98$.)

	n	$\bar{x}$	s
Class A	18	158.2	12.2
Class B	25	169.6	13.6

11. It is believed that a certain dosage of a specified drug will cause blood pressure to drop significantly. Test this hypothesis using the coded values of paired data shown below. (Use $\alpha = 0.05$.)

Patient	1	2	3	4	5	6	7	8	9	10
(B) Before drug	9	10	7	4	9	7	5	4	13	12
(A) After drug	5	7	7	2	10	5	2	3	11	11
$d = A - B$	-4	-3	0	-2	+1	-2	-3	-1	-2	-1

$\bar{d} = -1.7 \qquad s_d = 1.49$

12. Use a 95 percent confidence interval to estimate the mean increase in the cost of one liter of gas from August 17, 1981 to December 1, 1981. A random sample of 15 local gas stations' prices are listed below.

SELF-CORRECTING EXERCISES

Station	Aug. price	Dec. price
1	42.9	53.4
2	42.9	51.9
3	40.9	46.9
4	41.9	48.1
5	41.9	53.9
6	37.9	46.5
7	39.7	46.9
8	39.9	47.1
9	41.1	53.9
10	42.9	46.9
11	39.9	44.9
12	36.9	38.9
13	41.9	45.9
14	40.9	46.9
15	42.9	46.9

13. In a random sample of 400 married women, 372 responded no, while 358 out of a random sample of 400 married men responded no when asked, "Did your honeymoon live up to all your expectations." Test the hypothesis that there is no significant difference between the proportion of men and the proportion of women who responded no. (Use $\alpha = 0.01$.)

14. In a nation-wide sample of 600 school-age boys and 500 school-age girls, 288 boys and 125 girls admitted to having committed a destruction-of-property offense. Use this sample data to test the claim that boys are more likely to commit the destruction-of-property offense than are girls. (Use $\alpha = 0.10$.)

15. Worries about business conditions and the like seem to have caused the average family to save more than previously.
    a. Use a 95 percent confidence interval to estimate this proportion of increase based on the following sample. (p' is the percentage of families' income after taxes that was put into savings plans such as savings bank accounts, government bonds, and so forth.)

	Percentage saved	Number of families sampled
3rd quarter 1973	5.7	1,000
4th quarter 1973	6.0	1,200

    b. Does the confidence interval suggest a significant increase? Explain.

ADDITIONAL APPLICATIONS OF CHI-SQUARE

11

LESSON 11-1: THE MULTINOMIAL EXPERIMENT

A multinomial experiment is much like a binomial experiment, except that each individual trial can have any one of several outcomes (thus "multi-" instead of "bi-"). The specific number of different outcomes that are possible is identified by k. Each of the k possible outcomes has an expected probability p_i. In testing a multinomial experiment, we will compare the observed frequency with which each of the k outcomes occurs, O_i, to the frequency with which we had expected each outcome to occur, E_i. The value of each expected frequency will be found simply by multiplying the total number of trials, n, by the expected probability for that cell. ("Cells" are classes into which data are divided.)

$$E_i = n \cdot p_i \quad \text{for each } i = 1, 2, 3, \ldots, k$$

The sum of the E_i's, Σ_i, will be equal to the total number of trials n. (This should always be used as a check on your multiplications.) Also, be sure to observe the guideline that requires each of these expected values to be 5 or more, which may be accomplished by combining two or more cells of small expected values.

After the experiment is completed and each of the observed frequencies, O_i, has been determined, you will be able to calculate the test statistic. This test statistic will be the summation $\Sigma(O-E)^2/E$, where each of the k cells will contribute a numerical value, $(O-E)^2/E$. The total sum of addends over all cells will then become the test statistic we need. This summation has a χ^2 distribution. The complete formula, therefore, is:

$$\chi^{2*} = \sum_{\text{all cells}} \frac{(O-E)^2}{E} \qquad \text{[formula (11-1), p. 422 in textbook]}$$

This value is easiest to calculate by using a table. The following illustration shows the complete solution to a chi-square test of a multinomial experiment.

Illustration A class discussion about generating random numbers raised this question: Is it possible for people to act as a random-number generator just by naming

MULTINOMIAL EXPERIMENT   Lesson 11-1

numbers in some haphazard manner? To test this question, the class decided to ask each member of several different classes to name one single-digit number at random. One way to determine whether the numbers are random would be to test the frequency with which each digit was named — a multinomial experiment. If each digit was mentioned with approximately equal frequency, we can claim that each digit "had an equal chance," which is one important aspect of randomness. In the test, 200 students were asked to name a single-digit number with the following results.

digit	0	1	2	3	4	5	6	7	8	9
frequency	1	21	20	22	24	19	22	20	26	25

Test the claim that people name single-digit numbers with equal frequency, at $\alpha = 0.05$.

Solution
1. H_0: Each digit is named with equal frequency.
2. H_a: The digits are not all named with equal frequency.
3. (Figure 11-1.)

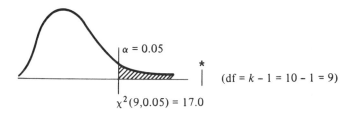

Figure 11-1

$\alpha = 0.05$

(df = k - 1 = 10 - 1 = 9)

$\chi^2(9, 0.05) = 17.0$

4. Each expected value is 20 (equal frequency means $p_i = 1/10$). (See Table 11-1.)

$$E_i = n \cdot p_i; \quad E_i = (200)(1/10) = 20$$

Table 11-1
Table for calculating χ^2 *

Digit	Observed	Expected	(O - E)	$\frac{(O - E)^2}{E}$
0	1	20	-19	361/20 = 18.05
1	21	20	1	1/20 = 0.05
2	20	20	0	0/20 = 0.00
3	22	20	2	4/20 = 0.20
4	24	20	4	16/20 = 0.80
5	19	20	-1	1/20 = 0.05
6	22	20	2	4/20 = 0.20
7	20	20	0	0/20 = 0.00
8	26	20	6	36/20 = 1.80
9	25	20	5	25/20 = 1.25
Totals	200	200		22.40

$$\chi^{2*} = \underline{22.40}$$

Note: The observed values, O, and the cell frequencies, are actually the same numbers.

5. Decision: Reject H_0.

Conclusion. It appears that people do not name all single-digit numbers with equal frequency. In fact, it seems safe to conclude that they discriminate against "zero," since the others seem to add very little to the total for chi-square.

The claim that each number is mentioned with equal frequency is not the only aspect of randomness: order of occurrence, odd-even relationship, and so forth, would also be of interest. But the example above does serve as an illustration of multinomial — perhaps your class has even been involved in a similar discussion.

EXERCISES 11-1

1. Another aspect of randomness in the illustration of the single-digit numbers might be the frequency of odd and even numbers. Using chi-square and the multinomial experiment, test the claim that the digits named were equally split between odd and even. (Use $\alpha = 0.05$.)

2. Draw a sample of 100 random digits from the random-number table in the text and test their randomness by testing:
    a. the equal frequency of all ten digits.
    b. the equal frequency of odd and even digits.

LESSON 11-2: THE CONTINGENCY TABLE

A contingency table is a two-way classification of enumerative data (counts) that is used to test the independence of two factors (or variables). A table of r rows and c columns is partitioned, and each of the n subjects is classified into one of the resulting cells. The contingency table shows the frequencies for each of these cells. The following illustration will amplify this concept.

Illustration Last semester 97 students studied statistics in classes taught by 4 different instructors. Using the data shown in Table 11-2, test the claim that the student's grade is independent of the instructor who taught his section. Use $\alpha = 0.05$.

CONTINGENCY TABLE   Lesson 11-2

Table 11-2

Instructor	A	B	C	D	F	
I	2	3	9	4	4	22
II	3	5	14	3	1	26
III	4	5	9	4	3	25
IV	6	6	10	1	1	24
	15	19	42	12	9	97

Solution   As the data are now grouped, the expected values for many of the cells are less than 5. (Recall that one of the guidelines for the chi-square tests says to avoid this situation.) Therefore, let's regroup the data into three categories of grade received: (1) above average (A or B), (2) average (C), and (3) below average (D or F); see Tables 11-3 and 11-4. Now only one expected value will be below 5, and it is nearly 5; therefore its value should not upset the test results.

Notes: 1. Table 11-3 shows the "observed values," which may be of any size.
2. Table 11-4 shows the "expected values" in parentheses; each of these values is to be 5 or larger.

Table 11-3

Instructor	A or B	C	D or F	
I	5	9	8	22
II	8	14	4	26
III	9	9	7	25
IV	12	10	2	24
	34	42	21	97

1. H_0: Final grade is independent of instructor.
2. H_a: Final grade is not independent of instructor.
3. (Figure 11-2.)

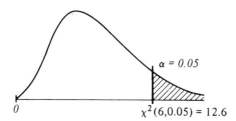

Figure 11-2 $\chi^2(6, 0.05) = 12.6$, $\alpha = 0.05$

4. Each $E_{ij} = (T_{i.} \times T_{.j})/T_{..}$, the product of the corresponding column and row totals divided by the grand total. These values are shown in parentheses in each cell in Table 11-4.

Table 11-4

	A or B	C	D or F	Row totals
I	5 (7.7)	9 (9.5)	8 (4.8)	22 (22.0)
II	8 (9.1)	14 (11.3)	4 (5.6)	26 (26.0)
III	9 (8.8)	9 (10.8)	7 (5.4)	25 (25.0)
IV	12 (8.4)	10 (10.4)	2 (5.2)	24 (24.0)
Column totals	34 (34.0)	42 (42.0)	21 (21.0)	97 (97.0)

$$\chi^{2*} = \Sigma \frac{(O-E)^2}{E} = 0.947 + 0.133 + 0.005 + 1.543 + \ldots = 8.65$$

(8.65 is less than 12.6.)

5. Decision: Fail to reject H_0.

Conclusion. There is no evidence found that would indicate the grade was affected by the instructor.

Let's look back at this example and try to understand the justification for using some of the formulas, in particular the formula for $E_{i,j}$. Each expected value for a cell is the result of multiplying the probability of that cell by the total sample size. For instance, in our example, we will need to find the probability that an element falls into the cell which is in row i and in column j. Then the probability, the proportion of the population that is in the cell, is multiplied by the sample size. Question: How do we find the probability for each of these cells?

Recall that the multiplication rule said that $P(A \text{ and } B) = P(A) \cdot P(B)$, provided A and B are independent. Therefore, the probability of a cell in the ith row and in the jth column, $P(i \text{ and } j)$, is equal to $P(i\text{th row})$ times $P(j\text{th column})$, provided the rows and columns can be considered independent. The independence can be assumed since it is the null hypothesis. Now the problem becomes, How do we obtain the probabilities for each of the rows and columns? The best we can do here is to estimate these probabilities by the use of the marginal totals. Thus the probability for column 9 will be estimated by the ratio of the total of column 1 to the total sample size. $P(\text{row } 1)$ is estimated by $P'(\text{row } 1) = T_{i=1}/T_{..}$. All the row and column probabilities can be estimated similarly. The probability for the cell in the ith row and the jth column is estimated by the product of the two corresponding marginal probabilities: $P'(\text{cell}_{i,j}) = P'(\text{row } i) \cdot$

P' (column j). The corresponding expected value is then found by the previously mentioned product.

$$E_{i,j} = T_{..} \times P' \text{ (cell }_{i,j})$$
$$= T_{..} \times P' \text{ (row } i) \times P' \text{ (column } j)$$
$$= T_{..} \times \frac{T_{i.}}{T_{..}} \times \frac{T_{.j}}{T_{..}}$$

$$E_{i,j} = \frac{T_{i.} \times T_{.j}}{T_{..}}$$

Once more we have seen one of our basic probability rules play an important role in the solution of a statistical problem.

EXERCISES 11-2

1. A psychology professor who was interested in investigating various types of "response bias" asked his students to predict the outcome of a coin toss to be performed in class. He categorized the responses as shown in Table 11-5.

Table 11-5

	Predicted		
Student	Head	Tail	
Male	32	35	67
Female	42	17	59
	74	52	126

Test at a 10 percent level of significance to see if the response and the sex of the respondent are independent.

SELF-CORRECTING EXERCISES FOR CHAPTER 11

1. Examine each of the following situations and determine whether or not it fits the form of a chi-square test. Explain your decision.
    a. Five hundred people are classified according to their blood type, and the number of persons in each category is recorded.
    b. A sample of 150 clothing garments is randomly selected from one day's production and inspected. Each garment is classified as being acceptable, irregular, or rejected, and the number in each category is recorded.
    c. The students enrolled in a large lecture class are classified according to

156 APPLICATIONS OF CHI-SQUARE Chapter 11

both their major academic field and their grade-point average. The number of students fitting each category is recorded.
d. The books in a home library are assigned to one of four categories: art, fiction, science, or reference. The number of books belonging to each category is recorded.

2. a. What are the characteristics of a multinomial experiment?
b. What restrictions must be observed when carrying out a multinomial experiment?
c. What are the characteristics of a contingency-table test?
d. What restrictions must be observed when carrying out a contingency-table test?

3. The number of felony arrests made per week by a large city police department were as follows: 12, 9, 20, 3, 12, 10, 15, 7, 8, 4. Do these frequencies support the belief that the number of arrests made per week was constant over this 10-week period? (Use $\alpha = 0.025$.)

4. Seventy-five nursery school children were asked to choose between three kinds of candy bar for a recess snack.

Kind of candy bar	Plain milk chocolate	Chocolate-covered soft nougat	Chocolate-covered crunchy caramel
Number of children	15	29	31

Test the hypothesis that the three kinds of candy bar are equally popular with nursery school children.
a. State the specific null hypothesis to be tested.
b. Complete the test using $\alpha = 0.01$.

5. In the genetic makeup of Four O'Clock flowers, the red gene(R) and the white gene(W) are both nondominant. Thus the resulting flower can be red, white, or pink depending on the cross that occurred. RR will result in a pure red flower, WW will result in a pure white flower, while RW and WR will both yield a pink flower. The chart below shows the possible results from crossing two pink flowers.

		2nd flower	
		R	W
1st flower	R	RR	RW
	W	WR	WW

To test the nondominant claim above, 500 crosses were made. The results are shown below.

SELF-CORRECTING EXERCISES

Color	Red	Pink	White	Total
Observed	112	271	117	500

Complete the hypothesis test above, using $\alpha = 0.05$.

Additional information: The genetic makeup of the resulting flower is such that it receives one gene from each flower being crossed. Each of the parent flower's genes has an equal chance to be received in the makeup of the resulting flower. That is: P(R from 1st flower) = $1/2$, P(W from 1st flower) = $1/2$, P(R from 2nd flower) = $1/2$, P(W from 2nd flower) = $1/2$.

Hint: the first chart above suggests a contingency table; however, due to the effect of crossing, two of the possible crosses produce the same result (pink). Therefore, you will need to use one row of three cells as indicated by the data.

6. A survey of 500 families with four children revealed the following distribution of male and female makeup.

Number of male and female children	4m 0f	3m 1f	2m 2f	1m 3f	0m 4f
Number of families	25	107	197	143	28

a. Explain why this is a binomial distribution.
b. Using this distribution of data, test the hypothesis that male and female births are equally likely. (Use $\alpha = 0.05$.)

7. A psychological experiment asks subjects to identify their preference among five choices: A, B, C, D, E. Four hundred and fifty-six people were asked to make their choices. The choices made are shown below, classified by male or female subject.

Choice	A	B	C	D	E
Male	60	33	51	32	0
Female	126	75	48	16	15

Test the claim that the person's sex did not in fact have an effect upon the preference. (Use $\alpha = 0.01$.)

8. The following chart shows the number of students who passed and who failed in four different instructors' classes for the same course.

Instructor	W	X	Y	Z	Total
Passed	51	46	63	38	198
Failed	7	15	10	16	48
Total	58	61	73	54	246

Test the hypothesis that the same proportion of students is passed by all four instructors.
a. State the specific null hypothesis to be tested.
b. Complete the test using $\alpha = 0.05$.

9. The same departmental final exam was taken by students enrolled in Calculus I during each of four successive years. Test the hypothesis that the distribution of grades during each of those four years was the same. Use $\alpha = 0.01$.

	Year			
Grade	1	2	3	4
Above average (A or B)	38	28	32	18
Average (C)	18	23	21	24
Below average (D or F)	22	29	12	24
	78	80	65	66

10. The following question has been asked regularly by the Gallup Poll for many years: "In politics, as of today, do you consider yourself a Republican, a Democrat, or an Independent?" Below are the results for four different years.

	Rep.	Dem.	Indep.
Oct. 1950	33	45	22
Nov. 1960	30	47	23
Oct. 1970	29	45	26
Jan. 1974	24	42	34

(These were percentage values — consider them as results of samples of size 100.)
a. State the appropriate null hypothesis and test it at $\alpha = 0.10$.
b. Using the data in the above chart as percentages, assume that 200 adults were asked in each year. Does this change in sample size (but not in percentage) have an effect? Recalculate χ^{2*} now and compare the results.
c. What do you think would happen if we were to double the sample size for each year, again, but keep the same percentages?

ANALYSIS OF VARIANCE

12

LESSON 12-1: THE ANOVA CONCEPT

The central idea of the ANOVA technique is the partitioning of the sum of squares to compare the amount of variation due to each of the possible sources of variation within the data. To partition a numerical value is to break it into two or more parts in such a way that the sum of the parts is equal to the whole. (22 could be partitioned into two parts, 20 and 2, and there are many other possibilities.) A good analogy to this would be the structure of a house. The total house is partitioned into rooms, and all the rooms together make up the house.

To see how the sum of squares is partitioned and how this is related to previous studies, let's look at an illustration. The illustration has been restricted to two levels so that we can compare the ANOVA techniques to the t test for the difference between two independent means.

Illustration Two samples, each of size 5, were obtained from each of two factor levels.

	\multicolumn{5}{c}{Replicates}				
Level A	2	3	4	2	4
Level B	5	6	7	6	6

Using this set of data let's test the null hypothesis that the mean level of A is the same as the mean level of B (the factor, as measured by levels A and B, does not affect the output). We'll test this hypothesis against a two-sided alternative, using $\alpha = 0.05$.

Solution I A t test for the difference between two independent means could be performed.

Sample							Σ
Sample A	x	2	3	4	2	4	15
	x^2	4	9	16	4	16	49
Sample B	x	5	6	7	6	6	30
	x^2	25	36	49	36	36	182

$$\bar{x}_A = \frac{15}{5} = 3.0 \qquad\qquad \bar{x}_B = \frac{30}{5} = 6.0$$

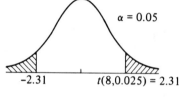

$$s_A = 1.0 \qquad\qquad s_B = 0.707$$

$$H_0: \mu_A - \mu_B = 0$$
$$H_a: \mu_A - \mu_B \neq 0$$

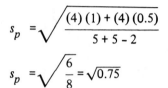

$\alpha = 0.05$

$-2.31 \qquad t(8, 0.025) = 2.31$

$$s_p = \sqrt{\frac{(4)(1) + (4)(0.5)}{5 + 5 - 2}}$$

$$s_p = \sqrt{\frac{6}{8}} = \sqrt{0.75}$$

$$t = \frac{3.0 - 6.0}{\sqrt{0.75}\,\sqrt{(1/5) + (1/5)}}$$

$$t = \frac{-3.0}{(0.866)(0.62)} = -5.2$$

Decision: Reject H_0.

Conclusion: The mean levels are different. The factor being tested does have an effect.

Solution II The ANOVA procedure could be followed to obtain an answer to the question. Typically with a problem this simple, the t test as shown above would be used. However, we want to show you how the ANOVA procedure works.

$H_0: \sigma^2_{Levels}/\sigma^2_{replicates} \leq 1$ (factor has no significant effect)
$H_a: \sigma^2_{Levels}/\sigma^2_{replicates} > 1$ (factor has a significant effect)

The null hypothesis that is actually being tested is the claim that the amount of variance between factor levels is no more than the amount of variance within the levels. This is a substitute for the claim that the mean levels of the tested factor are all the same, which is equivalent to "the variance between the means of the levels is zero, $\sigma_{\mu_r}^2 = 0$."

ANOVA CONCEPT   Lesson 12-1

$F(df_L, df_{rep}, 0.05) = F(1, 8, 0.05) = 5.32$ with $\alpha = 0.05$

$$SS(total) = \Sigma(x^2) - \frac{(\Sigma x)^2}{n} = [4 + 9 + 16 + \ldots + 36] - \frac{[2 + 3 + 4 + \ldots 6]^2}{10}$$

$$= 231 - \frac{(45)^2}{10} = 231.0 - 202.5 = \underline{28.5}$$

$$SS(factor) = \Sigma\left[\frac{C_i^2}{k_i}\right] - \frac{(\Sigma x)^2}{n} = \left[\frac{15^2}{5} + \frac{30^2}{5}\right] - \frac{(45)^2}{10}$$

$$= 225 - 202.5 = \underline{22.5}$$

$$SS(error) = \Sigma(x^2) - \Sigma\left[\frac{C_i^2}{k_i}\right] = 231 - 225 = \underline{6.0}$$

ANOVA Table

Source	SS	df	MS
Factor	22.5	1	22.50
Error	6.0	8	0.75
Total	28.5	9	

$F^* = \dfrac{22.50}{0.75} = 30.0$

$df(factor) = r - 1 = 2 - 1 = \underline{1}$

$df(error) = r(c - 1) = 2(5 - 1) = \underline{8}$

$df(total) = n - 1 = 10 - 1 = \underline{9}$

Decision: Reject H_0.

Conclusion: The factor being tested does have a significant effect.

We reached the same conclusion both ways. Now let's look at some of the specific points.

1. The "sum of squares" is the numerator in the formula for the variance of

data by definition $[\Sigma(x - \bar{x})^2]$. This sum may be found by use of the formula (see Exercise 1 of this lesson): $\Sigma x^2 - [(\Sigma x)^2/n]$.

The "sum of squares for total," SS(total), is this numerator when the total set of data is treated as one sample. If we take all ten pieces of data, we will find 28.5, just as we did in solution II. Why? The above formula is exactly the same as the one used in solution II, namely, formula (12-2).

2. The "sum of squares for error" is the same as the corresponding numerator for the pooled estimate of the variance within the samples. $s_p = \sqrt{6/8}$, numerator of 6. The SS(error) is also 6. This will always happen. The ss(error) is a measure of the variation within the rows. The pooled estimate of s, s_p, is an estimate that results from using the total sum of squares from within each level.

3. SS(factor) is, then, what is left.

$$\text{SS(total)} - \text{SS(error)} = \left[\Sigma(x^2) - \frac{(\Sigma x)^2}{n}\right] - \left[\Sigma(x^2) - \Sigma\left(\frac{c_i^2}{k_i}\right)\right]$$

$$= \Sigma\left[\frac{c_i^2}{k_i}\right] - \frac{(\Sigma x)^2}{n}$$

In effect, we are saying that any of the sums of squares for total that are not contained in the sum of squares for error must be due to the effect caused by the various levels of the test factor.

4. The mean square of error, MS(error), is the same as $(s_p)^2$, the pooled estimate of the variance of data within the levels.

5. MS(factor) is then the estimate for the variance between the means of the various levels of the factor tested.

6. MS(factor)/MS(error) is therefore a ratio of variances, and if this ratio is significant (falls in the critical region), then we say that there is evidence to cause us to reject the null hypothesis.

EXERCISES 12-1

1. Show that $\Sigma(x - \bar{x})^2 = \Sigma x^2 - (\Sigma x)^2/n$. Hint: Use the "definition" and the "shortcut" formulas for variance.

2. Consider the following table of data. It shows 4 replicates at each of three levels of a controllable test factor.

	Replicates			
Levels of factor	1	2	3	4
A	3	2	4	3
B	5	6	5	4
C	8	7	6	7

a. Calculate the sum of squares for total, error, and factor.
b. Calculate the three degrees of freedom and the two mean squares to complete the table.
c. Consider all 12 pieces of data and calculate the sum of squares using $\Sigma(x - \bar{x})^2$. Compare this answer to SS(total) found in part a.
d. Find a pooled estimate for the variance within the levels. Hint: A logical extension of the formula for s_p would be helpful. Compare this calculation to SS(error) and MS(error) found above.
e. Complete the F test at $\alpha = 0.05$.

SELF-CORRECTING EXERCISES FOR CHAPTER 12

1. Consider the following table of data for a single-factor ANOVA test.

Factor Levels

Replicates	I	II	III	IV	V
1	0	1	4	6	1
2	3	0	1	5	7
3	2	8	1	2	0

a. Find each of these individual values: $x_{1,1}, x_{1,3}, x_{2,1}, x_{2,4}, x_{3,5}, x_{4,6}, x_{3,1}, x_{4,2}$.

b. Find each of the totals: $C_1, C_2, C_3, C_4, C_5, T$.

c. Find each of these summations: $\Sigma x, \Sigma(x^2)$.

d. Determine the values of k, n, c.

2. Consider the following table of data for a single-factor ANOVA test.

Factor Levels

Replicates	I	II	III	IV
1	7	4	9	0
2	0	0	1	2
3	3	1	0	1
4	1	5	1	1
5	0	0	6	1
6	1	1	1	1

a. Find each of these individual values: $x_{1,1}, x_{1,3}, x_{2,1}, x_{2,4}, x_{3,5}, x_{4,6}, x_{3,1}, x_{4,2}$.

b. Find each of the totals: C_1, C_3, C_4, T.

c. Find each of the following sums: $\Sigma x, \Sigma(x^2)$.

d. Determine the values of k, n, c.

ANALYSIS OF VARIANCE Chapter 12

3. Use the data shown on the chart in Exercise 1 of this set of exercises.
   a. Calculate SS(total), SS(factor), and SS(error).
   b. Calculate df(total), df(factor), and df(error).
   c. Calculate MS(factor) and MS(error).
   d. Calculate F^*, [MS(factor)/MS(error)].

4. Use the data shown on the chart in Exercise 2 of this set of exercises.
   a. Calculate SS(total), SS(factor), and SS(error).
   b. Calculate df(total), df(factor), and df(error).
   c. Calculate MS(factor) and MS(error).
   d. Calculate F^*, [MS(factor)/MS(error)].

5. Let's look at three "made-up" illustrations.

	A: Factor Levels				B: Factor Levels				C: Factor Levels		
Replicates	I	II	III	Replicates	I	II	III	Replicates	I	II	III
1	2	4	3	1	3	5	2	1	2	1	4
2	3	3	5	2	3	5	2	2	4	3	6
3	4	2	4	3	3	5	2	3	3	2	5
4	5	5	2	4	3	5	2	4	5	4	3

   a. Calculate SS(total), SS(factor), and SS(error) for Illustration A.
   b. Calculate SS(total), SS(factor), and SS(error) for Illustration B.
   c. Do the answers found in parts a and b suggest that all the variation in A is within the row, whereas all the variation in B is between the rows? Explain.
   d. Calculate SS(total), SS(factor), and SS(error) for Illustration C.
   e. Suppose the amount of data were to double. What effect would this have on the calculated values, SS(total), SS(factor), and SS(error)?
   f. Carry out the calculation of SS(total), SS(factor) and SS(error) if the data were as shown in the chart below.

	Factor Levels		
Replicates	I	II	III
1	2	1	4
2	2	1	4
3	4	3	6
4	4	3	6
5	3	2	5
6	3	2	5
7	5	4	3
8	5	4	3

SELF-CORRECTING EXERCISES

g. Carry out the calculation of SS(total), SS(factor), and SS(error) if the data were as shown in the chart below.

	Factor Levels					
Replicates	I	II	III	IV	V	VI
1	2	2	1	1	4	4
2	4	4	3	3	6	6
3	3	3	2	2	5	5
4	5	5	4	4	3	3

h. Did the increase in data affect the calculated sum of squares as you expected?

6.

	Factor Levels		
Replicates	I	II	III
1	47	46	49
2	49	48	51
3	48	47	50
4	50	49	48

a. Use the data above and calculate SS(total), SS(factor), and SS(error).
b. Compare your answers in part a to the answer obtained in Exercise 5d.
c. Compare the data above to that in Illustration C and explain why the answers in part a are the same as those in answer 5d.

(This exercise suggests that we could use a coding technique whenever the set of data is too large — a coding technique which allows only for subtraction of a fixed value from each of the pieces of data.)

7. In an attempt to evaluate the effect of using high-test gasoline as opposed to regular, the following set of data (mpg per tank) was collected by an owner.

	Factor Levels	
Replicates	Regular	High-test
1	17.0	21.5
2	18.4	18.7
3	19.2	20.8

a. State the null hypothesis that will actually be tested using ANOVA on this data.
b. Interpret the meaning of this null hypothesis.
c. Complete the ANOVA test using $\alpha = 0.05$.

8.

Treatment

Replicates	A	B	C
1	10	4	7
2	12	10	4
3	9	9	5
4	14	6	4
5	10	4	8

The table of data above shows five observed values of a response variable for each of three different treatments.
a. State the null hypothesis to be tested.
b. Complete the testing of this null hypothesis using the ANOVA technique and $\alpha = 0.01$.

9. Four groups of students were selected, one group from each of four different classes, and tested at the end of a unit that all had studied. Does the following set of data suggest that all these students learned the same amount, or did a different amount of learning take place in each class?
Test Scores:

Factor Levels

Replicates	Class A	Class B	Class C	Class D
1	75	94	65	78
2	69	80	74	62
3	81	91	59	67
4	79	89	79	78
5	72	88	73	71
6	70		60	

a. State the null hypothesis that is to be tested using an ANOVA test.
b. Complete the ANOVA test using $\alpha = 0.05$.
Hint: Code the test scores before doing the calculations.

10. The following set of data shows the time required to run 220 yards on each of three different outdoor tracks. (20 has been subtracted from each entry.)

Track

Replicates	Cinder	Asphalt I	Asphalt II
1	7.9	6.8	6.7
2	7.5	5.5	6.0
3	7.7	7.4	5.3
4	7.9	7.4	7.1
5	7.9	5.1	7.1

a. What is the mathematical model for each x_{cr}?
b. Is there a significant difference in time due to the track? (Use $\alpha = 0.01$.)

…

LINEAR CORRELATION AND REGRESSION ANALYSIS

13

LESSON 13-1: THE RANDOM ERROR ϵ

For linear bivariate data, each piece of data (x,y) fits the mathematical model $y = \beta_0 + \beta_1 x + \epsilon$. The $(\beta_0 + \beta_1 x)$ part of the equation can be thought of as the mean value of all the y-values that correspond to that x-value. Thus the value of ϵ ("epsilon," the random error) is the difference between the observed value of y and the mean value of all y's that correspond to that same x.

$$\epsilon = y - (\beta_0 + \beta_1 x)$$

We will, in practice, not know the true values of β_0 and β_1, the true population values for the y-intercept and the slope, nor will we know the mean value of y for each of the possible x-values. Thus, the best substitute we have available for estimating ϵ will be $e = y - (b_0 + b_1 x)$, or $e = y - \hat{y}$ • $\hat{y}$ becomes our estimate for the mean value of y at each of the x-values: $\hat{y} = b_0 + b_1 x$. The values of b_0 and b_1 are the coefficients for the line of best fit as calculated from the observed data using the "least-squares criterion." The least-squares criterion requires that $\Sigma(y - \hat{y})^2$ be as small as possible. If this is the case, then it must follow that $\Sigma(y - \hat{y})$ will be equal to zero, implying that $\hat{y}$ has at least some of the properties of a mean. Recall that the sum, $\Sigma(x - \bar{x})$, is always equal to zero for any given sample.

Illustration Find the amount of error, e, present in the ordered pair $(4,7)$ where $\hat{y} = 1.0 + 0.5x$.

Solution The amount of error in $(4,7)$ must be the amount of error in the 7. At $x = 4$, $\hat{y} = 1.0 + 0.5(4) = 3.0$, therefore the value of $y - \hat{y}$, the estimate of the error, is $+4.0$ ($e = y - \hat{y} = 7.0 - 3.0 = 4.0$).

Now that you have seen what the random error e is and how its measure is estimated, we are ready to proceed to the next lesson and study the variance of this random error.

EXERCISES 13-1

1. Find the estimate for the amount of random error present in the y-values of

each of the following ordered pairs if the line of best fit is $\hat{y} = 1.0 + 0.5x$.
  a. (4,5)
  b. (6,3)
  c. (8,5)

LESSON 13-2: VARIANCE ABOUT THE LINE OF BEST FIT

The variance of a variable was defined back in Chapter 2. That definition might now be restated as: The sum of the squares of the deviations from the mean divided by the appropriate number of degrees of freedom. The variance of a variable y would then be $\Sigma(y - \bar{y})^2/df$. However, with bivariate data, as the value of x changes, the mean value of y follows the line of best fit. Therefore, the above formula would become:

$$s_e^2 = \frac{\Sigma(y - \hat{y})^2}{n - 2}$$ [formula (13-6), p. 488 in textbook]

The number of degrees of freedom for this situation is defined in the textbook as $n - 2$. Formula (13-6) is then rewritten as:

$$s_e^2 = \frac{(\Sigma y^2) - (b_0)(\Sigma y) - (b_1)(\Sigma xy)}{n - 2}$$ [formula (13-8), p. 489 in textbook]

The following illustration shows the value of s_e^2 calculated by using the above formulas in a manner related to previous calculations of variance. $y - \hat{y}$ is the error, and if all the errors are squared and summed together, we have the value of the numerator for formulas (13-6) and (13-8). This numerator is represented by the symbol SSE, "*the sum of squares for error*" (which you recall from Chapter 12).

Illustration Consider the following set of 16 ordered pairs of x- and y-values.

x	2	2	4	4	5	6	6	6	8	8	9	10	10	13	13
y	1	3	2	4	3.5	2	3	5	6	3	7	5.5	5	7	6.5	8.5

$\Sigma x = 112$ $\Sigma y = 72$ $\Sigma x^2 = 956$ $\Sigma xy = 590$ $\Sigma y^2 = 393$

The calculated line of best fit is $\hat{y} = 1.0 + 0.5x$, as shown in Figure 13-1.

VARIANCE ABOUT LINE OF BEST FIT   Lesson 13-2

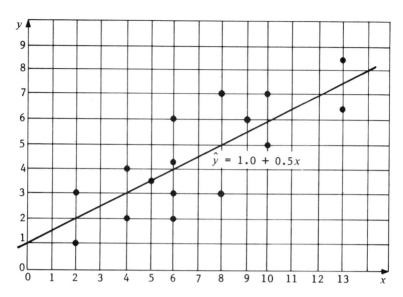

Figure 13-1

The observed estimates for the random error, e [shown in the same order as points (x,y) appear above] are:

$$e = -1, +1, -1, +1, 0, -2, -1, +1, +2, -2, +2, 0, -1, +1, -1, +1$$

(See Figure 13-2.)

The variance of e will be found by two methods: first by use of formula (13-8), the formula that you will ordinarily use. SSE = $(\Sigma y^2) - (b_0)(\Sigma y)$

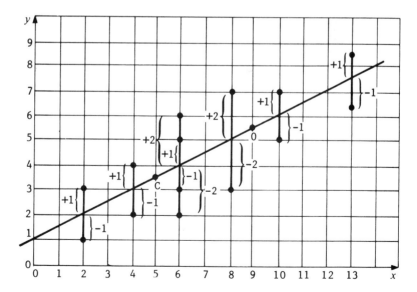

Figure 13-2

$-(b_1)(\Sigma xy) = 393 - (1.0)(72) - (0.5)(590) = 393 - 72 - 295 = 26$. Second, the value of SSE can be found by $\Sigma(y - \hat{y})^2$, which is $\Sigma(e)^2 = (-1)^2 + (+1)^2 + (-1)^2 + \ldots + (-1)^2 + (+1)^2 = 26$. The variance of the y-values about the line of best fit then is $s_e^2 = \text{SSE}/(n - 2) = 26/14 = 1.86$.

As you see, the very same value for SSE, the numerator of the variance formula, results from both methods of calculation. The use of formula (13-6) was practical for this illustration since the set of data was made up for this purpose. It is also a familiar form for the calculation of the variance. In data obtained as a result of sampling, the use of formula (13-8) will generally be more practical in terms of the ease of arithmetic.

The value s_e^2, then, represents the measure of variability of the data about the line of best fit in the same manner that s^2, as defined and calculated in Chapter 2, represented the measure of the variability or spread of single-variable data about its mean. This calculated value of s_e^2, then, is a measure which serves as an estimate of the true value of the variance of y about the true line of best fit within any linearly related bivariate data.

EXERCISES 13-2

1. Given the following 18 pairs of data.

x	1	2	2	3	3	3	4	4	4	4	5	5	6	6	6	7	7	8
y	4.5	3	5	2	3.5	5	1	2	4	5	1.5	3.5	1	2	3	0.5	2.5	1

a. Draw a scatter diagram of this data.
b. Find the equation of the line of best fit and draw it on the scatter diagram.
c. Calculate s_e^2:
   (1) using formula (13-6) and the 18 observed values of e_i.
   (2) using formula (13-8) and the results of part b.

LESSON 13-3: CONFIDENCE-INTERVAL ESTIMATES FOR REGRESSION

The two confidence-interval estimates made in connection with regression are the interval estimate for the mean value of y at a given value of x, $\mu_{y|x_0}$, and the interval estimate for the value of an individual y at a given value of x, y_{x_0}. The general form for both of these confidence-interval estimates is the same as used

CONFIDENCE-INTERVAL ESTIMATES Lesson 13-3

previously for a normally or an approximately normally distributed random variable, namely:

$$\text{statistic} \pm t(df, \alpha/2) \cdot \text{standard error of the statistic}$$

The point estimate for both $\mu_{y|x_0}$ and y_{x_0} is the value of $\hat{y}$ at x_0. This value of $\hat{y}$ serves as the "statistic" in the general form shown above.

The standard error is the lone remaining factor to be determined. Each of the two confidence intervals will have a different standard error of estimate. We are estimating a mean value in one case and individual values in the other. Individual values of y are more variable than are mean values, thus the confidence interval for individual values will be considerably wider than the corresponding interval for a mean value.

Before we look at the formulas for the two standard errors, let's look back at a formula for the standard error used in a previous case. In particular, let's look at the standard error for the difference between two independent means, $(\sigma_{\bar{x}_1 - \bar{x}_2})^2$. σ_1 and σ_2 are the standard deviations for each of the two populations, and the formula for $\sigma_{\bar{x}_1 - \bar{x}_2}$ was given as

$$\sqrt{\frac{\sigma_1^2}{n_1} + \frac{\sigma_2^2}{n_2}}$$

Consider for a moment the variance of one sample mean alone. The central limit theorem stated that the standard deviation of $\bar{x}$'s was equal to $\sigma/\sqrt{n}$, therefore the variances of $\bar{x}$'s, $\sigma_{\bar{x}}^2$, must be σ^2/n. Therefore, we can say that the variance of $\bar{x}_1$ is σ_1^2/n_1 and the variance $\bar{x}_2$ is σ_2^2/n_2. The formula for the standard error for the difference of two independent means tells us to add these two variances to obtain the variance for the difference in the two means. The standard error, then, is the square root of this sum. The point we are getting at is that the variances from each of these two sources, the two separate sampling distributions, are added together to obtain the variance present in the combined situation. This is exactly what happens when the standard error of estimate for our regression intervals is formed.

When the mean value of y at a given x is being estimated, we have two sources of variance involved, two sampling distributions if you wish. One source of variance is the sampling distribution of means itself. The other is the sampling distribution of observed value for the slope. The variance for the distribution of means is estimated by s_e^2/n. The variance for the distribution of slopes is estimated by $s_e^2/\Sigma(x - \bar{x})^2$. Inspection of Figure 13-11 on page 454 in the textbook will convince you that the effect of this slope variance is magnified as the value of x_0 moves away from the mean value of x; in fact, at $\bar{x}$, there is no effect due to the slope. A "correction factor" is used in connection with variance due to the slope-sampling distribution. This factor is the square of the distance from the x_0

under study to the value of $\bar{x}$. Thus we add these two variance estimates together to obtain the variance of mean values of y about the line of best fit. The square root of the variance is taken and the result is:

$$\text{standard error of estimate} = \sqrt{\frac{s_e^2}{n} + (x_0 - \bar{x})^2 \cdot \frac{s_e^2}{\Sigma(x - \bar{x})^2}}$$

This can be simplified by factoring the s_e^2 out from under the radical, leaving us with:

$$s_e \sqrt{\frac{1}{n} + \frac{(x_0 - \bar{x})^2}{\Sigma(x - \bar{x})^2}}$$

The insertion of this quantity into the basic form suggested previously will give us formula (13-14).

$$\hat{y} \pm t(n - 2, \alpha/2) \cdot s_e \cdot \sqrt{\frac{1}{n} + \frac{(x_0 - \bar{x})^2}{\Sigma(x - \bar{x})^2}} \qquad \text{[formula (13-14), p. 498 in textbook]}$$

As we have previously seen, the term $\Sigma(x - \bar{x})^2$ is difficult to calculate and can be replaced with $[n(\Sigma x^2) - (\Sigma x)^2]/n$. The use of this expression in formula (13-4) brings about formula (13-15) which is in more usable form:

$$\hat{y} \pm t(n - 2, \alpha/2) \cdot s_e \cdot \sqrt{\frac{1}{n} + \frac{n(x_0 - \bar{x})^2}{n(\Sigma x^2) - (\Sigma x)^2}} \qquad \text{[formula (13-15), p. 498 in textbook]}$$

Illustration Using the set of 16 ordered pairs and the results obtained in the illustration in Lesson 13-2, construct the 90 percent confidence interval for the mean value of y at $x = 10$.

Solution Information from before:

$\Sigma x = 112 \qquad \Sigma x^2 = 956 \qquad n = 16 \qquad s_e^2 = 1.86$

$\hat{y} = 1.0 + 0.5x$

$\bar{x} = \dfrac{\Sigma x}{n} = \dfrac{112}{16} = 7.0$

$\hat{y} = 1.0 + 0.5(10) = 1.0 + 5.0 = 6.0$

Using formula (13-15):

$$6.0 \pm t(14, 0.05) \cdot \sqrt{1.86} \cdot \sqrt{\frac{1}{16} + \frac{16(10 - 7)^2}{16(956) - (112)^2}}$$

CONFIDENCE-INTERVAL ESTIMATES   Lesson 13-3

$$6.0 \pm (1.76)(1.363)\sqrt{0.0625 + \frac{144}{2752}}$$

$$6.0 \pm (1.76)(1.363)\sqrt{0.0625 + 0.0523}$$

$$6.0 \pm (1.76)(1.363)\sqrt{0.1148}$$

$$6.0 \pm 0.81$$

$5.19 < \mu_{y|x=10} < 6.81$ (90 percent confidence interval)

This interval is shown on the diagram in Figure 13-3.

The standard error of estimate for an individual value of y at a given x contains one more term. In addition to the variance of the mean value and the variance of the slope, we must also account for the variance among the individual values of y. Thus the variance among individual values of y at a given x will be estimated by the sum of s_e^2/n and $(x - \bar{x})^2 \cdot s_e^2/[\Sigma(x - \bar{x})^2]$ along with s_e^2, the variance of the individuals about the line of best fit. This variance, then, is estimated by:

$$s_e^2 + \frac{s_e^2}{n} + (x - \bar{x})^2 \cdot \frac{s_e^2}{\Sigma(x - \bar{x})^2}$$

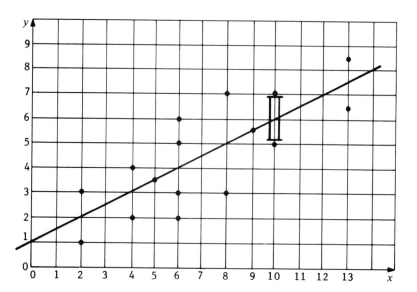

Figure 13-3

The estimated value for the standard error of estimate then becomes:

$$s_e \sqrt{1 + \frac{1}{n} + \frac{(x_0 - \bar{x})^2}{\Sigma(x - \bar{x})^2}}$$

LINEAR CORRELATION AND REGRESSION   Chapter 13

As before we will replace the $\Sigma(x - \bar{x})^2$, so the resulting formula for the estimation of an individual value of y becomes:

$$y \pm t(n - 2, \alpha/2) \cdot s_e \cdot \sqrt{1 + \frac{1}{n} + \frac{n(x_0 - \bar{x})^2}{n(\Sigma x^2) - (\Sigma x)^2}}$$

Illustration Construct the 90 percent confidence-interval estimate for an individual value of y at $x = 10$, using the illustration from Lesson 13-2.

Solution Most of the information and work is the same as found in the previous illustration. The only difference is that the number under the radical is one larger than in the previous problem: Therefore, the interval becomes:

$6.0 \pm (1.76)(1.363)\sqrt{1.1148}$

6.0 ± 2.53

$3.47 < y_{x=10} < 8.53$

This interval is shown on the diagram in Figure 13-4. You might compare it to the interval found previously.

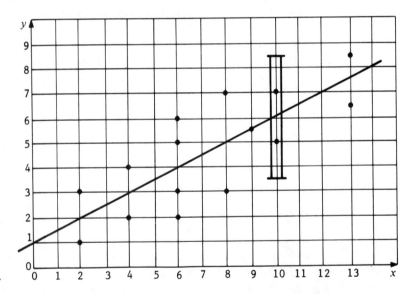

Figure 13-4

EXERCISES 13-3

1. For the illustration used in this lesson, construct the 90 percent confidence interval for the following.
   a. the mean value of y at $x = 7$
   b. an individual value of y at $x = 7$

2. Using the data and the information obtained in Exercise 1 of Lesson 13-2, construct the 95 percent confidence-interval estimate for the following.
   a. the mean value of y at $x = 3$
   b. the mean value of y at $x = 7$
   c. an individual value of y at $x = 3$
   d. an individual value of y at $x = 7$

SELF-CORRECTING EXERCISES FOR CHAPTER 13

1. Consider the following set of ordered pairs.

x	0	1	1	2	2	3	3	4
y	2	2	3	2	4	3	4	4

   Note: All the odd-numbered items in this exercise set use this set of data.

   a. Plot these eight ordered pairs to form a scatter diagram. (Use a whole sheet of graph paper for your scatter diagram, since several of the following exercises require you to draw on the same graph.)
   b. Calculate the mean value of the x's, $\bar{x}$, and the mean value of the y's, $\bar{y}$.
   c. Locate the point $(\bar{x},\bar{y})$ on your scatter diagram. Draw a vertical and horizontal line through $(\bar{x},\bar{y})$.
   d. Calculate the covariance of x and y.

2. Consider the following set of ordered pairs.

x	1	2	3	3	4	5	6	7	7	8
y	5	4	2	4	3	3	2	1	3	1

   Note: All the even-numbered items in this exercise set use this set of data.

   a. Plot these 10 ordered pairs on a coordinate axis to form a scatter diagram. (Use a whole sheet of graph paper for your scatter diagram, since several of the following exercises require you to draw on the same graph.)
   b. Calculate $\bar{x}$ and $\bar{y}$ (to the nearest tenth) and draw the lines on the scatter diagram that represent these values.
   c. Calculate the covariance of x and y (to the nearest hundredth).

3. Use the set of eight ordered pairs of data from Exercise 1 above.
   a. Calculate the standard deviation of x, s_x, and the standard deviation of y, s_y. (These are two separate but similar problems.)
   b. Calculate the linear correlation coefficient using formula (13-2) and the

necessary information previously found in answers to Exercises 1 and 3a.
c. Calculate the linear correlation coefficient using formula (13-3).
d. Compare the answers found in parts b and c.

4. Use the set of 10 pairs of data given in Exercise 2 above.
   a. Calculate the standard deviation of both x and y, s_x and s_y.
   b. Calculate the linear correlation coefficient using formula (13-2) and the previously found answers from Exercises 2 and 4a.
   c. Calculate the linear correlation coefficient using formula (13-3).
   d. Did you obtain the same answer in parts b and c?

5. a. Test the null hypothesis, $\rho = 0$, against the alternative, rho is positive ($\rho > 0$) with respect to the data in Exercise 1. (Use $\alpha = 0.05$.)
   b. Determine the 95 percent confidence-interval estimate for the true value of the population linear correlation coefficient.

6. a. Complete a two-tailed hypothesis test of $\rho = 0$ on the data given in Exercise 2. (Use $\alpha = 0.05$.)
   b. Determine the 95 percent confidence-interval estimate for the true value of ρ.

7. Use the set of data given in Exercise 1.
   a. Calculate the equation of the line of best fit, $\hat{y} = b_0 + b_1 x$.
   b. Using the equation found in part a, calculate the value of $\hat{y}$ that corresponds to each of the different values of x. ($x = 0, 1, 2, 3, 4$.)
   c. Find each of the eight values of e, the difference between the observed and the expected value of y at each value of x.
   d. Using formula (13-6), calculate the value of s_e^2, the variance of y about the line of best fit.
   e. Calculate the value of s_e^2 using the alternative formula (13-8).
   f. Do your answers in parts d and e agree?

8. Use the set of data given in Exercise 2.
   a. Calculate the equation of the line of best fit, $\hat{y} = b_0 + b_1 x$. (Round b_0 and b_1 to the nearest hundredth.)
   b. Using the equation found in part a, calculate the value of y that corresponds to each of the values of x. ($x = 1, 2, 3, \ldots, 8$.)
   c. Find the 10 values of e that represent the difference between the 10 observed values of y and their corresponding expected values ($\hat{y}$).
   d. Calculate the variance about the line of best fit, s_e^2, using formula (13-6), and then calculate it again using formula (13-8).
   e. Compare the results of the two different calculations in part d.

9. Use the necessary information from the odd-numbered exercises above.
   a. Calculate the standard error of estimate for b_1, s_{b_1}.
   b. Test the null hypothesis of $\beta_1 = 0$ against the one-sided alternative that β_1 is positive. (Use $\alpha = 0.05$ and clearly state your conclusion.)

SELF-CORRECTING EXERCISES

c. Construct the 95 percent confidence-interval estimate of the true value of slope, β_1.
d. On the scatter diagram drawn in answer to Exercise 1, draw lines through the point $(\bar{x}, \bar{y})$ that represent the two extreme values found in the confidence interval in part c.

10. Use the necessary information from the even-numbered exercises above.
    a. Calculate the standard error of estimation for b_1, namely s_{b_1}.
    b. Test the null hypothesis $\beta_1 = 0$ against the two-sided alternative that β_1 is different from zero. (Use $\alpha = 0.05$ and clearly state your conclusion.)
    c. Construct the 95 percent confidence-interval estimate of the true value of slope, β_1.
    d. On the scatter diagram drawn in answer to Exercise 2, draw lines through the point $(\bar{x}, \bar{y})$ that represent the two extreme values found in the confidence interval in part c.

11. a. Calculate the 95 percent confidence-interval estimate for the mean value of y at $x = 1$ ($u_{y|x=1}$) for the data given in Exercise 1 using formula (13-15).
    b. Calculate the 95 percent confidence-interval estimate for the mean value of y at $x = 0, 2, 3,$ and 4.
    c. Draw a vertical line segment on the scatter diagram for this data representing each of the five confidence intervals above.
    d. Draw the confidence-interval belts on the same scatter diagram and then use your graph to find the 95 percent confidence-interval estimate for the mean value of y at $x = 3.5$.

12. a. Calculate the 95 percent confidence-interval estimate for an individual value of y at $x = 3$ ($y_{x=3}$) for the data given in Exercise 2 using formula (13-16).
    b. Calculate the 95 percent confidence-interval estimate for individual values of y at $x = 1, 5,$ and 7.
    c. Draw a vertical line segment on the scatter diagram representing each of the four confidence intervals above.
    d. Draw the confidence-interval belts on your scatter diagram. Use your graph to make the 95 percent confidence-interval estimate for an individual y at $x = 4$.

ELEMENTS OF NONPARAMETRIC STATISTICS

14

LESSON 14-1: THE CONCEPT OF NONPARAMETRICS, DEMONSTRATED WITH THE SIGN TEST

Nonparametric tests revolve around relatively elementary probability theory. The sign test, which we will use to illustrate this fact, depends upon the binomial probability distribution. The binomial experiment as applied here has two outcomes, a "plus" or a "minus," and they are hypothesized to be equally likely. Therefore, p and q will both be assigned the value $1/2$, and the binomial random variable, x, will be the number of the less-frequent sign found in the observed data. (It's easier to count the less-frequent signs since there are fewer of them.) The number of trials, n, will be the total number of plus and minus signs observed; zeroes will not be counted at all.

The null hypothesis for the sign test (as well as for many of the other nonparametrics) typically claims that the two samples are from populations having the same distribution. This then implies that their means, their variances, and so forth are the same. If this is, in fact, the case, then the expected observed relative frequencies of + and − differences would be $1/2$. If the observed data supports this, we will want to accept the "same distribution" claim. If it is not the case that the observed frequencies agree, then we will want to reject the claim of "same distribution."

Let's inspect a specific situation and demonstrate the use of the table of critical values and the working of the hypothesis test. Let's consider a situation where a sample contains 15 signs. The null hypothesis is equivalent to $P(+) = P(-) = 0.5$; therefore, if H_0 is true, as we assume while testing, then the probability of any specific number of plus signs is the value shown in Table 14-1.

Table 14-1

$x =$	0	1	2	3	4	5	6	7
$P(x) =$	.0+	.0+	.003	.014	.042	.092	.153	.196
$x =$	8	9	10	11	12	13	14	15
$P(x) =$	.196	.153	.092	.042	.014	.003	.0+	.0+

The table of critical values for the sign test, Table 11 on page 584 in the textbook, shows a critical value of 3 for a two-tailed test at $\alpha = 0.05$. The critical value found in the table is the "maximum number of the least frequent sign that will allow us to reject H_0." Thus for our above situation, we could "reject H_0" if the observed value of x is 0, 1, 2, or 3. Notice that the probabilities for $x = 15$, 14, 13, and 12 are exactly the same as for $x = 0, 1, 2,$ and 3. This is why we will always count the least frequent sign. The sum of the actual probabilities for all critical values (all values in the critical region) will need to be equal to or less than $\alpha/2$. (Our test is two tailed and the other half of α is on the other tail.) For $n = 15$, the probability of the critical region is then $P(x = 0, 1, 2, 3) = P(x = 0) + P(x = 1) + P(x = 2) + P(x = 3) = (0.0+) + (0.0+) + (0.003) + (0.014) = 0.017$. Therefore, α is actually $2(0.017) = 0.034$. If $x = 4$ (and therefore $x = 11$) were considered to be part of the critical region, then α would jump to 0.118, and this is larger than the desired value for α. Thus the value of x that would cause α to be exactly 0.05 would be somewhere between the 3 and the 4. However, such a value is meaningless since x is a count and must be a whole number. Thus we settle for the 3 as a critical value, and interpret $\alpha = 0.05$ as meaning "no more than a 5 percent" chance of committing the type I error. Notice also from the above information that the critical value for $\alpha = 0.10$ and $n = 15$ is also 3. Why?

Note: We added the probabilities in the above paragraph. Why is this justified?

Recall that guidelines in Chapter 5 told us that x was approximately normally distributed with a mean of np and a standard deviation of $\sqrt{npq}$ wherever n was greater than 20 and both np and nq were larger than 5. Both of these guidelines are met whenever n is greater than 20 in the sign test; thus the sign test may be carried out by using z as a test statistic. There is, however, one adjustment that must be made. x is a discrete variable, while the normal distribution is a distribution of a continuous variable. The adjustment is as simple as adding (or subtracting) the value 1/2 to the number x. The following diagram (Figure 14-1) will show this concept. Recall, x is discrete; therefore the probability for a given

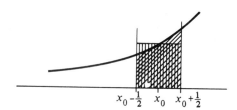

Figure 14-1

Vertical shaded area represents $P(x = x_0)$, the binomial probability; diagonal shaded area represents $P(x_0 - \frac{1}{2} < x < x_0 + \frac{1}{2})$, the estimate using normal distributions.

x, $P(x = x_0)$, is approximated by the measure of $P(x_0 - 1/2 < x < x_0 + 1/2)$, as shown above. As described above, the interval from $(x_0 - 1/2)$ to $(x_0 + 1/2)$ must be completely in the critical region in order to reject the null hypothesis. Therefore, x', the adjusted random variable, is equal to $x + 1/2$ if x is smaller than $n/2$, and is equal to $x - 1/2$ when x is larger than $n/2$, as x_1 and x_2 show on the figure below (Figure 14-2).

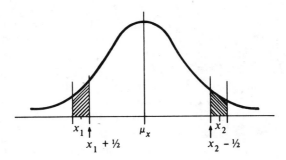

Figure 14-2

The calculation of z follows our familiar format:

$$z = \frac{\text{statistic} - \text{parameter}}{\text{standard deviation of statistic}}$$

$$z = \frac{x' - \mu}{\sigma_x} = \frac{x' - n/2}{\sqrt{n}/2}$$

[formula (14-1), p. 519 in textbook]

$n/2$ is the mean value of the random variable x. $\mu_x = np$ for the binomial distribution. $\sigma_x = \sqrt{npq} = \sqrt{(n)(1/2)(1/2)} = \sqrt{n}/2$.

EXERCISES 14-1

1. Explain why the critical value for $n = 15$ and $\alpha = 0.10$ is the same as for $n = 15$ and $\alpha = 0.05$.

2. Explain why the addition of probabilities, $P(x = 0, 1, 2, 3) = P(x = 0) + P(x = 1) + P(x = 2) + P(x = 3)$ on the previous page was justifiable.

3. A sign test is to be completed, where $n = 30$ and $\alpha = 0.05$.
   a. If x is the number of the least-frequent sign in a two-tailed test, what values of x fall in the critical region?
   b. If x is the number of "plus" signs in a two-tailed test, what values of x fall in the critical region?

4. Find the critical value of x, the number of the least-frequent sign in a two-tailed sign test, where $n = 50$ and $\alpha = 0.15$.

LESSON 14-2: THE MANN-WHITNEY U TEST

The Mann-Whitney U test makes use of the ranking of data, whereas the sign test merely considers whether data is above or below a certain median value. The U test enables us to determine whether two independent samples have been selected from the same population. The Mann-Whitney U test is the non-parametric alternative to the t-test for the difference of two means and is used when certain assumptions for parametric testing are not met.

Illustration A softball fan claims there is no difference in the number of runs scored during softball games played during daylight hours or played under artificial light. To support the claim, the fan obtains a random sample of runs scored in 8 games under each lighting condition. Based on the data shown in Table 14-2, test the fan's claim at $\alpha = 0.05$.

Table 14-2

Game condition	Runs scored								
day	19	9	13	8	6	18	16	14	25	21
night	17	19	11	18	9	14	20	18	22	23

Solution I A U test is used to test the null hypothesis that the run-scoring abilities are the same. We use $\alpha = 0.05$.

Step 1. H_0: There is no difference in run-scoring abilities in day or night games.

H_a: There is a difference in run-scoring abilities in day or night games.

Rank the data into one sample. A rank of 1 is assigned to the team scoring the least runs (day or night) and a rank of 20 is given to the team scoring the most runs. See Table 14-3.

Step 2. Determine the critical value for U (Table 12 in Appendix E). For $n_a = n_b = 10$ and at $\alpha = 0.05$ in a two-tailed test, the critical value is 23.

Step 3. Calculate the test statistic U. The Mann-Whitney U statistic is defined by using the following pair of formulas:

$$U_a = n_a \cdot n_b + \frac{(n_b)(n_b + 1)}{2} - R_b \qquad \text{[formula (14-3), p. 525 in textbook]}$$

$$U_b = n_a \cdot n_b + \frac{(n_a)(n_a + 1)}{2} - R_a \qquad \text{[formula (4-4), p. 525 in textbook]}$$

Table 14-3

Ranked data	Rank before adjustment	Rank after adjustment	Lighting condition	Ranks for days	Ranks for nights
6	1	1	day	1	
8	2	2	day	2	
9	3 ⎤ $\frac{3+4}{2} = 3.5$	3.5	day	3.5	
9	4 ⎦	3.5	night		3.5
11	5	5	night		5
13	6 ⎤ $\frac{6+7}{2} = 6.5$	6.5	night		6.5
13	7 ⎦	6.5	day	6.5	
14	8	8	day	8	
16	9	9	night		9
17	10	10	day	10	
18	11 ⎤	12	day	12	
18	12 ⎬ $\frac{11+12+13}{3} = 12$	12	night		12
18	13 ⎦	12	night		12
19	14 ⎤ $\frac{14+15}{2} = 14.5$	14.5	day	14.5	
19	15 ⎦	14.5	night		14.5
20	16	16	night		16
21	17	17	day	17	
22	18	18	night		18
23	19	19	night		19
25	20	20	day	20	
Sum				94.5	115.5

where

n_a = size of sample a R_a = sum of ranks for sample a

n_b = size of sample b R_b = sum of ranks for sample b

and U, the test statistic, will be the smaller of U_a and U_b.

In our problem we have (see Table 14-3):

$n_a = n_{day} = 10$ $R_a = R_{day} = 94.5$

$n_b = n_{night} = 10$ $R_b = R_{night} = 115.5$

Then we have:

$$U_a = U_{day} = (10)(10) + \frac{(10)(11)}{2} - 115.5 = 39.5$$

$$U_b = U_{night} = (10)(10) + \frac{(10)(11)}{2} - 94.5 = 60.5$$

Therefore:

$$U = \underline{39.5}$$

MANN-WHITNEY U TEST   Lesson 14-2

Step 4. Decision: Since $U = 39.5$ and the critical value is 23 and $23 < 39.5$, we fail to reject the null hypothesis.

Conclusion: The fan has no evidence to indicate that there is a difference in run-scoring ability due to lighting conditions.

Solution II   Let us work the softball problem again, this time using the test statistic z. We can use this approach because the distribution of U is approximately normal. Step 1 is the same as in Solution I.

Step 2. For $\alpha = 0.05$ and for a two-tailed test, the critical region is as shown in Figure 14-3.

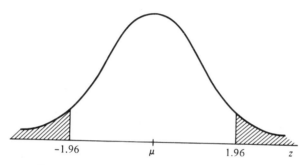

Figure 14-3

Step 4. Determine the mean and standard deviation of the U's, using these formulas:

$$\mu_U = \frac{n_a \cdot n_b}{2} \qquad \text{[formula (14-5), p. 526 in textbook]}$$

$$\sigma_U = \sqrt{\frac{n_a n_b (n_a + n_b + 1)}{12}} \qquad \text{[formula (14-6), p. 526 in textbook]}$$

Then we have:

$$\mu_U = \frac{10 \cdot 10}{2} = 50$$

$$\sigma_U = \sqrt{\frac{10 \cdot 10(10 + 10 + 1)}{12}} = 13.23$$

Now we calculate a z-value, using this formula:

$$z = \frac{U - \mu_U}{\sigma_U} \qquad \text{[formula (14-7), p. 526 in textbook]}$$

$$z^* = \frac{39.5 - 50}{13.23} = \underline{0.79}$$

Step 4. Decision: Since the test statistic does not fall in the critical region ($-0.79 > -1.96$ and $-0.79 < 1.96$), we fail to reject H_0.

Conclusion: There is no evidence to indicate that there is a difference in run-scoring abilities.

EXERCISES 14-2

1. A VW owner decides to test the merits of switching to a better grade of gasoline. He keeps a record of the miles/gallon he obtained when using two brands of gasoline. Do the following data indicate that he gets better mileage from the better grade of gasoline? Test at $\alpha = 0.05$, using the Mann-Whitney U test.

Gasoline	Miles/gallon
regular	37 41 25 45 43 21 36 44 26 30 41
supreme	47 43 32 40 31 29 21 33 26 30 37

LESSON 14-3: THE RUNS TEST

We are suspicious of the randomness of data that have too few or too many runs. A *run* is a sequence of data displaying a common property. When the property is gone, a new run begins. We will use the runs test to test the null hypothesis of randomness. It will be rejected if the number of runs is too small or too large. The assumption made using the one-sample runs test is that each observation is classified as one type or another. The procedure for completing the hypothesis test is as follows:

Step 1. H_0: The sequence is random.

H_a: The sequence is not random.

Step 2. First determine the number of each classification (n_1, n_2). Then determine the test statistic and the critical values and critical region according to these criteria:

Case I: If n_1 and n_2 are both equal to or less than 20, and a two-tailed test is desired, then the number of runs is the test statistic and Table 13 in Appendix D of the textbook is used to determine critical values.

Case II. If either n_1 or n_2 (or both) are larger than 20 or if $\alpha \neq 0.05$, the test statistic z will be used.

Step 3. Determine the number of runs V in the sample. If case II is involved, use formulas (14-8), (14-9), and (14-10) to calculate the observed value of z.

RUNS TEST Lesson 14-3

Step 4. Compare the results of step 4 with the critical value(s) from step 3. Make a decision and write a conclusion based on this comparison.

Illustration Let us demonstrate the runs test with the following problem: Cash drawer shortages in a large department store are under investigation. A manager, who suspects that part-time employees are making more errors than full-time employees, collected the names of 9 clerks to determine whether they are part-time (P) or full-time (F) help and then recorded the following sequence of cash shortages:

$$F, P, P, P, F, P, P, P, F$$

At $\alpha = 0.05$, do these data show sufficient lack of randomness to confirm the manager's suspicion?

Solution Step 1. H_0: The sequence is random.

H_a: The sequence is not random.

Step 2. Find the critical values of V. $\alpha = 0.05$ and

n_1 = number of full-time employees = 3

n_2 = number of part-time employees = 6

From Table 13 of Appendix E in the textbook, the critical values are 2 and 8.

The critical region includes the values of V less than or equal to 2 or greater than or equal to 8.

Step 3. There are 5 runs ($V^* = 5$):

F PPP F PPP F

Step 4. Decision: Fail to reject H_0 (5 is in the noncritical region).
Conclusion: There is no evidence to contradict the null hypothesis that the shortages are random.

Illustration A fair coin is tossed 44 times and produces 23 heads and 21 tails. There is a question about whether the coin flips were random events. Test the hypothesis that the tosses were random. (Use $\alpha = 0.05$.) The data occurred as follows:

HHHTTTHHTTHHHHTTTHHHHHHHTTHTTHHHTTTTHHHTTTT

Solution Step 1. H_0: The sequence is random.

H_a: The sequence is not random.

Step 2. $n_1 = 23$ and $n_2 = 21$. Since $n_1 > 20$ and $n_2 > 20$, the z statistic will be used.

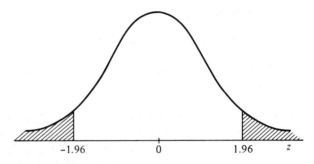

Step 3. The runs are

HHH TTT HH TT HHHH TTT HHHHHHH TT H TT HHH TTTTT HHH TTTT

So the number of runs is $V = 14$.

Since n_1 and n_2 are larger than 20, the observed value of z will be calculated using formulas (14-8), (14-9), and (14-10).

$$\mu_V = \frac{2(21)(23)}{21+23} + 1 = 22.95$$

$$\sigma_V = \sqrt{\frac{2(23)(21)(2 \cdot 23 \cdot 21 - 23 - 21)}{(23+21)^2(23+21-1)}} = 3.27$$

$$z^* = \frac{14 - 22.95}{3.27} = -2.74$$

Step 4. Decision: We reject H_0. ($z^* = -2.74$ is in the critical region.)
Conclusion: The sequence is not random.

EXERCISES 14-3

1. A statistics professor, during the five-day school week, sometimes eats at the faculty dining room and other times eats a sandwich at home. The following data were reported (F = faculty dining room, H = home) about the sequence of the professor's eating habits:

   FFFHFHFHFFFHFFFFHFHHHFFFFHFF

   Test the professor's claim, at $\alpha = 0.05$, that her choice of dining sites is random.

2. A radio talk show host claims there is a lack of randomness in the number of

men and women that call his show. The following sequence of calls was received during his two-hour Monday morning show:

WWWMMWMMM

Test the announcer's claim at $\alpha = 0.05$.

LESSON 14-4: SPEARMAN RANK CORRELATION COEFFICIENT

In Chapters 3 and 13, the correlation coefficient r was described as an index measuring the association between two variables. An alternative method of correlation is used in nonparametric statistics. It is called the *Spearman rank correlation coefficient* and is found by using this formula:

$$r_s = 1 - \frac{6\Sigma(d_i)^2}{n(n^2 - 1)}$$ [formula (14-11), p. 535 in textbook]

where

$$d_i = x_i - y_i$$

n = number of pairs of data

x_i, y_i = rankings of variables x, y

The procedure for calculating the Spearman rank correlation coefficient is as follows:
 1. Rank the x's and the y's separately.
 2. Establish ordered pairs of rank numbers by pairing the ranks for each ordered pair (x, y).
 3. Compute d_i ($d_i = \text{rank}_{x_i} - \text{rank}_{y_i}$) and d_i^2 for each ordered pair.
 4. Find the sum Σd_i^2.
 5. Calculate the value of r_s using formula (14-11) from page 575 of the textbook.

Note: If ties occur in the set of x-values or in the set of y-values, the rank assigned to each value involved in the tie is equal to the mean value of the positions that these tied data occupy. This idea is shown in the following illustration.

Illustration Ten students in a large class were randomly identified and the amount of time each needed to complete a quiz was noted. Their grades and amounts of time are listed below.

Student	A	B	C	D	E	F	G	H	I	J
Time, x	5	10	5	15	20	25	10	5	15	20
Grade, y	30	40	50	60	40	60	30	20	70	50

NONPARAMETRIC STATISTICS   Chapter 14

Calculate the value of r_s, Spearman's rank correlation coefficient.

Solution   Rank the 10 x-values; see Table 14-4.

Table 14-4

Time, x	Rank before adjustment	Rank after adjustment
5	1 ⎫	2
5	2 ⎬ $\frac{1+2+3}{3} = 2$	2
5	3 ⎭	2
10	4 ⎫ $\frac{4+5}{2} = 4.5$	4.5
10	5 ⎭	4.5
15	6 ⎫ $\frac{6+7}{2} = 6.5$	6.5
15	7 ⎭	6.5
20	8 ⎫ $\frac{8+9}{2} = 8.5$	8.5
20	9 ⎭	8.5
25	10	10

Rank the 10 y-values; see Table 14-5.

Table 14-5

Score, y	Rank before adjustment	Rank after adjustment
20	1	1
30	2 ⎫ $\frac{2+3}{2} = 2.5$	2.5
30	3 ⎭	2.5
40	4 ⎫ $\frac{4+5}{2} = 4.5$	4.5
40	5 ⎭	4.5
50	6 ⎫ $\frac{6+7}{2} = 6.5$	6.5
50	7 ⎭	6.5
60	8 ⎫ $\frac{8+9}{2} = 8.5$	8.5
60	9 ⎭	8.5
70	10	10

Now do the calculations for d_i; see Table 14-6.

Using formula (14-11), we obtain:

$$r_s = 1 - \frac{6(64)}{10(100-1)} = \underline{0.612}$$

Illustration   Does the sample in the illustration above provide sufficient evidence to conclude that there is a positive correlation between quiz scores and time? Use $\alpha = 0.05$.

Solution   Step 1. H_0: Scores and time are independent.

SPEARMAN RANK CORRELATION COEFFICIENT   Lesson 14-4

Table 14-6

i	Time, x_i	Score, y_i	Rank of x_i	Rank of y_i	$d_i =$ rank$_{x_i}$ − rank$_{y_i}$	d_i^2
1	5	30	2	2.5	−0.5	0.25
2	10	40	4.5	4.5	0.0	0.00
3	5	50	2	6.5	−4.5	20.25
4	15	60	6.5	8.5	−2.0	4.00
5	20	40	8.5	4.5	4.0	16.00
6	25	60	10	8.5	1.5	2.25
7	10	30	4.5	2.5	2.0	4.00
8	5	20	2	1	1.0	1.00
9	15	70	6.5	10	−3.5	12.25
10	20	50	8.5	6.5	2.0	4.00
Sum						64.00

H_a: Scores and time are positively correlated (longer times result in higher quiz scores).

Step 2. $n = 10$; find the critical value in Table 14 of Appendix E in the textbook.

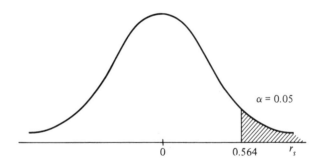

Step 3. r_s was calculated in the previous illustration; it is $r_s^* = 0.612$.

Step 4. Decision: Reject H_0. (0.612 falls in the critical region.)
Conclusion: There is sufficient evidence to support the contention that longer time spent completing the quiz results in a higher grade.

EXERCISES 14-4

1. The news media and the fans of a professional basketball team voted for the most valuable player. The votes were counted and the MVP rankings for each group were as listed below.

Player	Rank from fans	Rank from media
A	1	1
B	2	3
C	3	2
D	4	5
E	5	4

Is there a positive linear correlation between the rankings at the 5% level?

2. The yearly salaries of the basketball players in Exercise 1 are listed below. Is there a correlation between the news media's MVP rankings and the player's salaries? Test at $\alpha = 0.10$.

Player	Salary
A	$300,000
B	175,000
C	200,000
D	75,000
E	100,000

SELF-CORRECTING EXERCISES FOR CHAPTER 14

1. The sign test may be used as an alternative to the test for one population mean. Explain how the sign is arrived at and justify the test-statistic's interpretation.

2. Explain why the sign test is an alternative to a test of the difference between two dependent means and not an alternative for the difference of two independent means.

3. Explain the difference between Spearman's rank correlation coefficient and Pearson's linear correlation coefficient.

4. A random sample of 16 customers (numbered from 1 to 16) is asked to compare two brands of paper towels and comment on the absorption power. The following table shows the evaluations made by the customers.

SELF-CORRECTING EXERCISES

Customer stating product A better	Customer stating product B better	Customer stating no difference
1	3	5
2	4	8
14	6	11
	7	12
	9	13
	10	
	15	
	16	

Do the data support the claim that the customers prefer Product B to Product A? Use the sign test to test this alternative hypothesis at $\alpha = 0.05$.

5. Members of a consumer testing group are asked to indicate which of two proposed magazine advertisements they prefer. Do the following data indicate that Ad 2 is preferred to Ad 1? Use $\alpha = 0.05$ and the sign test to test the appropriate null hypothesis.

Consumer	Ad 1	Ad 2	Sign
1	yes		−
2		yes	+
3	yes		−
4		yes	+
5		yes	+
6		yes	+
7		yes	+
8		yes	+

6. Use the Mann-Whitney test to test the null hypothesis that the following two samples come from populations with equal average values. Use a level of significance of $\alpha = 0.10$.

A	5	8	9	3	5	6	3	3	9	8
B	4	9	9	2	8	5	7	2	7	6

7. A jogger keeps a record of the days he jogs (J) and does not jog (N) during a period of two months. He collected the following data:

JJJJNJNJJNNNNNNJJNJJNNJNJNNNJNJJJJNJJNJJNJNJJNNJNJJJJNJJJJJ

Do these data show a lack of randomness in the jogger's habits? Use $\alpha = 0.05$.

8. Eleven students are selected at random to test the hypothesis that SAT mathematics and verbal scores are independent. Their scores are listed below.

Student	SAT scores	
	Math	Verbal
1	450	390
2	440	350
3	460	450
4	450	450
5	480	380
6	450	520
7	470	360
8	470	390
9	450	420
10	560	440
11	400	430

a. Compute the Spearman rank correlation coefficient, r_s.
b. Is there sufficient evidence to fail to reject the null hypothesis that these scores are uncorrelated? Use $\alpha = 0.05$.

9. Two tests were given to 12 candidates for a typing position. Each candidate was tested for speed and then accuracy of typing. The twelve candidates are then ranked in each category, as shown in the following table.

Candidate	Rank typing speed	Accuracy
1	5	7
2	6	5
3	1	2
4	3.5	3
5	11	11
6	12	12
7	3.5	4
8	9.5	10
9	9.5	9
10	2	1
11	8	8
12	7	6

Is there a positive correlation between typing speed and accuracy? Test at $\alpha = 0.05$.

SOLUTIONS

INTRODUCTION

SOLUTIONS TO EXERCISES I-1

1. a. $2y$ b. y^2 c. $3y + 6$ d. $y^2 - 2$ e. $\sqrt{y} + 2$
2. a. II, III, IV   b. I–2; II–1; III–1
   c. 3   d. II–3, x, y; IV–$n, n-1$
3. a. the number 2 more than x, multiplied by 5
   b. two more than 5 times the number x
   c. the sum of x squared and y squared
   d. the square of the sum of x and y

SOLUTIONS TO EXERCISES I-2

1. a. 3   b. 10.5   c. 50.5
2. a. 50   b. 36
3. a. 3   b. –5   c. $2\frac{1}{3}$
4. a. 21   b. $3\frac{1}{2}$ c. 97   d. 441   e. 141
5. a. $\frac{1}{6}$ b. 1.0   c. $2\frac{1}{3}$ d. 6
6. a. 16   b. 20   c. 72   d. 40   e. 58
   f. 100   g. 36

SOLUTIONS TO EXERCISES I-3

1. 2625, 6196, 5469, 1302, 1874, 6667, 5904, 0154, 3909, 2501, 7485, 0545
2. 171, 043, 176, 207, 338, 106, 333, 341, 048, 113
3. 24
4. T, T, T, T, H, H, H, H, T, T: 4 heads, 6 tails
5. (1,5), (3,2), (1,5), (1,4), (1,4), (5,4), (5,3), (2,1), (2,4), (5,4)

SOLUTIONS TO EXERCISES 1-4

1.

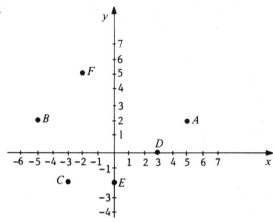

2. a.

Point	x	y
A	-2	7
B	0	5
C	1	4
D	3.5	1.5
E	6	-1

b.

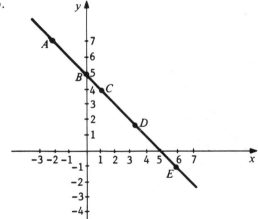

SOLUTIONS: Introduction

3. Pick any values of x you wish; $x = -2, 0, 1, 2, 4$ will be convenient.

x =	-2	0	1	2	4
y =	-7	-4	-2½	-1	2

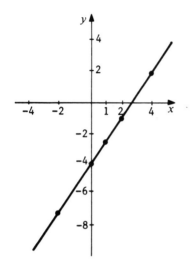

4. $m = 10$, $b = -3$, $\underline{y = 10x - 3}$

5. a.

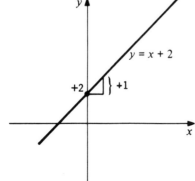

b.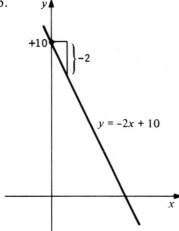

c.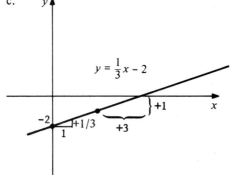

6. I. (a) Given points (3,1) and (9,5).

$$m = \frac{\Delta y}{\Delta x} = \frac{5-1}{9-3} = \frac{4}{6} = \frac{2}{3}$$

$y = \frac{2}{3}x + b$ [and passes through (3,1)]

$1 = \frac{2}{3}(3) + b$ implies that $b = -1$

Thus $y = \frac{2}{3}x - 1$.

SOLUTIONS: Introduction

I. (b)

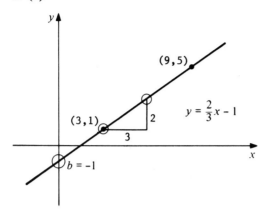

II. (a) Given points (-2,3) and (6,-1).

$$m = \frac{\Delta y}{\Delta x} = \frac{(-1)-(3)}{6-(-2)} = \frac{-4}{8} = -\frac{1}{2}$$

$$y = -\frac{1}{2}x + b \qquad \text{[use (-2,3)]}$$

$$3 = (-\frac{1}{2})(-2) + b \text{ implies } b = 2$$

Thus $y = -\frac{1}{2}x + 2$.

(b)

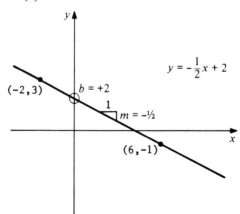

SOLUTIONS TO EXERCISES 1-5

1.
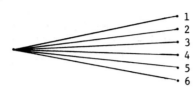

2.
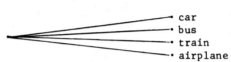

3. a.
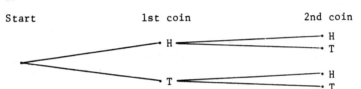

   b.
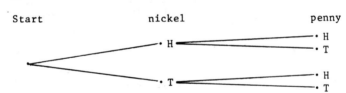

   c. no

4.
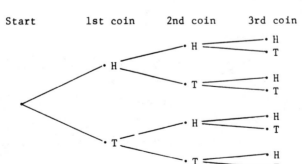

SOLUTIONS: Introduction

5. a.

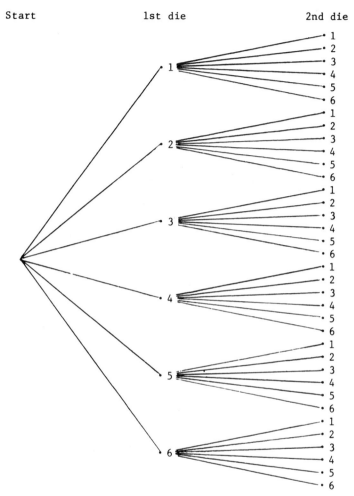

b. 36 branch ends

6.

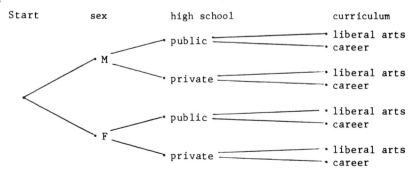

SOLUTIONS: Introduction

7. a.

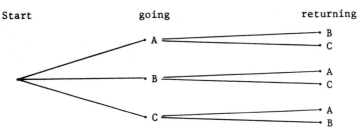

b. 6
c. 2

SOLUTIONS TO EXERCISES I-6

1. a. $B \cap C = \{1\}$ b. $\bar{B} = \{3,4,5,6\}$
   c. $B \cup C = \{0,1,2,3,5\}$ d. $B \cup \bar{C} = \{0,1,2,4,6\}$
   e. $\bar{B} \cup C = \{1,3,4,5,6\}$ f. $(\overline{B \cap C}) = \{0,2,3,4,5,6\}$
   g. $(\overline{B \cup C}) = \{4,6\}$
   h. $\bar{B} \cup \bar{C} = \{3,4,5,6\} \cup \{0,2,4,6\} = \{0,2,3,4,5,6\}$

2. a. Bio ∩ Stat = {Beth, Jim}
      These people are enrolled in both courses.
   b. Bio ∪ Stat = {Angela, Beth, Jim, Renee, Norm, Bob, Steve, Russ, Pat, Mack}
      There are 10 people altogether. Beth and Jim belong to both groups, but each count only once in the union.

3. a. The subset of art majors who are enrolled in statistics.
   b. The total combined set of the set of black and foreign students.
   c. The set of black veterans.
   d. The set of students made up of all the liberal arts majors and all those who have completed a statistics course.

SOLUTIONS TO EXERCISES I-7

1. a. 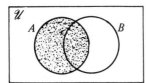   b.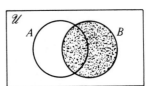

SOLUTIONS: Introduction

c.
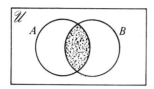

d.
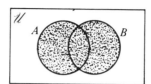

e.
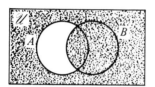

f.
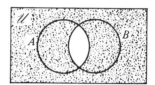

2. a.
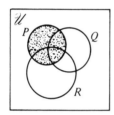

b.
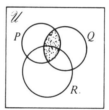

c.
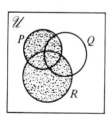

d.
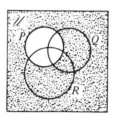

e.
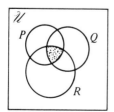

f.
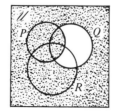

g.
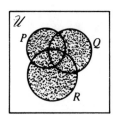

h.
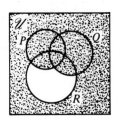

SOLUTIONS TO EXERCISES I-8

1. $4! = 4 \times 3 \times 2 \times 1 = \underline{24}$

2. $6! = 6 \times 5 \times 4 \times 3 \times 2 \times 1 = \underline{720}$

3. $8! = 8 \times 7 \times (6!) = 56 \times 720 = \underline{40320}$

4. $(6!)(8!) = (720)(40320) = \underline{29{,}030{,}400}$

5. $\dfrac{8!}{6!} = \dfrac{8 \times 7 \times (6!)}{6!} = 8 \times 7 = \underline{56}$

6. $\dfrac{8!}{4!\,4!} = \dfrac{8 \times 7 \times 6 \times 5 \times (4!)}{4 \times 3 \times 2 \times 1 \times (4!)} = 2 \times 7 \times 5 = \underline{70}$

7. $\dfrac{8!}{6!\,2!} = \dfrac{8 \times 7 \times 6!}{(6!) \times 2 \times 1} = 4 \times 7 = \underline{28}$

8. $2\left(\dfrac{8!}{5!}\right) = 2\left(\dfrac{8 \times 7 \times 6 \times 5!}{5!}\right) = 2 \times 8 \times 7 \times 6 = \underline{672}$

SOLUTIONS TO EXERCISES I-9

1. a. $\dfrac{5 \cdot 4 \cdot 3 \cdot 2 \cdot 1}{4 \cdot 3 \cdot 2 \cdot 1 \cdot 1} = 5$

   b. $\dfrac{6 \cdot 5 \cdot 4 \cdot 3 \cdot 2 \cdot 1}{3 \cdot 2 \cdot 1 \cdot 3 \cdot 2 \cdot 1} = 20$

   c. $\dfrac{10 \cdot 9 \cdot 8 \cdot 7 \cdot 6 \cdot 5 \cdot 4 \cdot 3 \cdot 2 \cdot 1}{4 \cdot 3 \cdot 2 \cdot 1 \cdot 6 \cdot 5 \cdot 4 \cdot 3 \cdot 2 \cdot 1} = 210$

SOLUTIONS: Chapter 1

d. $\dfrac{17 \cdot 16 \cdot 15 \cdot 14 \ldots 1}{15 \cdot 14 \ldots 1 \cdot 2 \cdot 1} = \dfrac{17 \cdot 16}{2 \cdot 1} = 136$

e. $\dfrac{8 \cdot 7 \cdot 6 \cdot 5 \cdot 4 \cdot 3 \cdot 2 \cdot 1}{5 \cdot 4 \cdot 3 \cdot 2 \cdot 1 \cdot 3 \cdot 2 \cdot 1} = 56$

2. $\binom{n}{k} = \dfrac{n!}{k!\,(n-k)!}$

$\binom{n}{n-k} = \dfrac{n!}{(n-k)!\,[n-(n-k)]!} = \dfrac{n!}{(n-k)!\,k!}$

They are the same.

3. They both represent a number of objects which must be able to be counted; therefore, they must be positive integers.

4. Table values: the answers are the same as in Exercise 1.

5. $\binom{21}{5} = \binom{20}{4} + \binom{20}{5} = 4845 + 15504 = 20349$

CHAPTER I

SOLUTIONS TO EXERCISES 1-1

1. a. Population: All members of the American work force.
   b. Variable: Do you fear losing your job? (Yes or no.)
   c. Sample: The workers who were polled.
   d. Parameter: The proportion of all workers who fear job loss. Reported to be 1/5.
   e. Statistic: The proportion of the polled workers who fear job loss.

2. a. Population: All members of the working force who are covered by state and federal unemployment insurance programs.
   b. Variable: Amount of annual salary.
   c. Parameter: Average amount of annual salary for all workers for each state.

SOLUTIONS TO EXERCISES 1-2

1. a. (C) An academic average would be a continuous variable since almost any decimal value might occur.

b. (B) The number of students per homeroom on the honor roll would be discrete. This variable represents a count of the number of honor students found in each element of the population (a homeroom).

c. (A) The data here are attribute. The population is the set of individual elementary school children and the response from each element is "yes" or "no" to the question of membership on the safety patrol. Thus each child has been classified as being a member or nonmember of the patrol. The sample statistic reported was the total number in the member classification.

Note: You should remember to identify the elements of the population and then ask yourself what information is asked for from each of these elements.

d. (C) A number of minutes is a length of time and thus is a measurement. Therefore it is a continuous variable.

e. (B) The population is the dozens of eggs that are on the grocery shelf. The variable is the count of the number of cracked eggs found in each dozen. Therefore it is a discrete variable.

f. (A) The population is the set of all people who bought new cars during 1973. Each person bought a "station wagon" or "something different from a station wagon." Thus this is a classification and the variable is attribute.

g. (B) The population consists of all the laboratory mice. The variable is the number of shocks received by each member, thus the variable is discrete (a count of the number of shocks).

SOLUTIONS TO SELF-CORRECTING EXERCISES FOR CHAPTER 1

1. a. Population: All members of the bar association.
   b. Sample: The 206 members who answered the questionnaire.
   c. Variable: Amount of annual salary.
   d. Parameter: Average annual salary for all members.
   e. Statistic: Average annual salary for the 206 members.
   f. Experiment: The questionnaire.

2. a. Population: All American boys.
   b. Variable: Amount of monthly allowance received.
   c. Parameter: Average monthly allowance for all American boys.

3. a. Population: All adult women.
   b. Variable: Number of children each had given birth to.
   c. Sample: The adult women polled to obtain the information that this article was based on.

d. Parameter: Average number of children given birth to for all women.
e. Statistic: Average number of children given birth to by the women sampled.

CHAPTER 2

SOLUTIONS TO EXERCISES 2-1

1. Use the mode to present the idea of most typical: mode = 1
   Use the mean to present the idea of the average number of people transported by each car:

   $$\bar{x} = \Sigma x/n = 50/25 = 2.0$$

2. Use the mean to present the idea of the average value of property involved:

   $$\bar{x} = 363/12 = 30.25$$

3. By definition the median is the value that falls exactly in the middle when the data is arranged in order according to size. Half of the values are larger and half are smaller. The "middle man" is often thought of as average.

SOLUTIONS TO EXERCISES 2-2

1. a. $\bar{x} = \Sigma x/n = 24/6 = 4.0$

x	$(x - \bar{x})$	$(x - \bar{x})^2$
7	3	9
2	-2	4
3	-1	1
4	0	0
3	-1	1
5	1	1
	0	16

$s = \sqrt{16/5}$

$s = \sqrt{3.2}$

$s = 1.8$

b. $\bar{x} = \Sigma x/n = 32/8 = 4.0$

x	$(x - \bar{x})$	$(x - \bar{x})^2$
1	-3	9
2	-2	4
5	1	1
8	4	16
4	0	0
3	-1	1
7	3	9
2	-2	4
	0	44

$s = \sqrt{44/7}$
$s = \sqrt{6.2857}$
$s = \underline{2.5}$

c. $\bar{x} = \Sigma x/n = 120/10 = 12.0$

x	$(x - \bar{x})$	$(x - \bar{x})^2$
10	-2	4
8	-4	16
11	-1	1
19	7	49
7	-5	25
14	2	4
12	0	0
16	4	16
12	0	0
11	-1	1
	0	116

$s = \sqrt{116/9}$
$s = \sqrt{12.88}$
$s = \underline{3.6}$

d. $\bar{x} = \Sigma x/n = 17/5 = 3.4$

x	$(x - \bar{x})$	$(x - \bar{x})^2$
2	-1.4	1.96
3	-0.4	0.16
4	0.6	0.36
3	-0.4	0.16
5	1.6	2.56
	0	5.2

$s = \sqrt{5.2/4}$
$s = \sqrt{1.3}$
$s = \underline{1.1}$

SOLUTIONS: Chapter 2

2. a.

x	x^2
7	49
5	25
8	64
9	81
7	49
36	268

$$s = \sqrt{\frac{\Sigma x^2 - \frac{(\Sigma x)^2}{n}}{(n-1)}}$$

$$s = \sqrt{\frac{268 - \frac{(36)^2}{5}}{(4)}}$$

$$s = \sqrt{2.2} = \underline{1.5}$$

b.

x	x^2
9	81
8	64
12	144
11	121
14	196
9	81
10	100
12	144
85	931

$$s = \sqrt{\frac{931 - \frac{(85)^2}{8}}{(7)}}$$

$$s = \sqrt{3.982}$$

$$s = \underline{2.0}$$

c.

x	x^2
4	16
3	9
2	4
9	81
6	36
7	49
5	25
6	36
4	16
5	25
51	297

$$s = \sqrt{\frac{297 - \frac{(51)^2}{10}}{(9)}}$$

$$s = \sqrt{4.10}$$

$$s = \underline{2.0}$$

SOLUTIONS TO EXERCISES 2-3

1. a. $n = 40, k = 10; nk/100 = (40)(10)/100 = 4.0$; therefore, $i = \underline{4.5}$
   b. $n = 40, k = 25; nk/100 = (40)(25)/100 = 10.0$; therefore, $i = \underline{10.5}$
   c. $n = 40, k = 38; nk/100 = (40)(38)/100 = 15.20$; therefore, $i = \underline{16}$

SOLUTIONS: Chapter 2

2. 1 2 3 4 5 . . . 48 49 50
   a. $i = (n + 1)/2 = (50 + 1)/2 = \underline{25.5}$
   b. $n = 50, k = 75; nk/100 = (50)(75)/100 = 37.5$; therefore, $i = \underline{38}$
   c. $i =$ 1   2   3 . . . 25 26 . . . 38 39 . . . 49 50
      _____
            (50) (49) (48) . . . (26) (25) . . . (13) (12) . . . (2) (1)
      For median: $i = (n + 1)/2 = (50 + 1)/2 = \underline{25.5}$ (halfway between the 25th and 26th values)
      For third quartile: $nk/100 = (50)(25)/100 = 12.5$; therefore, $i = \underline{13}$ (the 38th value from L and the 13th value from H are the same)

3. To find P: $k = 50; nk/100 = n(50)/100 = n/2$.
   Case 1: If n is even, $n/2$ is an integer and $i = n/2 + 1/2 = (n + 1)/2$.
   Case 2: If n is odd, $n/2$ contains the fraction one-half. Therefore, $i =$ next larger integer $= n/2 + 1/2 = (n + 1)/2$.
   Both cases result in $i = (n + 1)/2$.

4. 1 2 3 . . . 37 38 39 . . . 56 57 58 . . . 74 75
   a. $i = (n + 1)/2 = (75 + 1)/2 = \underline{38}$
   b. $n = 75, k = 75; nk/100 = (75)(75)/100 = 56.25$; therefore, $i = \underline{57}$
   c. 1   2   3 . . . 37 38 39 . . . 56 57 58 . . . 74 75
                             ↕   ↕
       (75) (74) (73) . . . (39) (38) (37) . . . (20) (19) (18) . . . (2) (1)
                             $\tilde{x}$ Q_3
      For median: $i = (n + 1)/2 = (75 + 1)/2 = 38$ from the highest-valued data.
      For third quartile: $nk/100 = (75)(25)/100 = 18.75$; therefore, $i = 19$ from the highest-valued data.
      As shown above, the results are the same as found in parts a and b.

SOLUTIONS TO EXERCISES 2-4

1.

x	f
5	2
10	1
15	5
20	13
25	10
30	5
35	2
40	1
45	1
Sum	40

SOLUTIONS: Chapter 2

2.

x	f
1	2
2	3
3	9
4	5
5	3
6	1
7	1
8	0
9	1
Sum	25

3.

Class limits	f
60–62	6
63–65	10
66–68	10
69–71	19
72–74	4
75–77	1
Sum	50

SOLUTIONS TO EXERCISES 2-5

1. a.

x	1	1	1	2	2	2	2	3	3	3	3	3	3	4	4	4	4	4	5	5
x^2	1	1	1	4	4	4	4	9	9	9	9	9	9	16	16	16	16	16	25	25

$\Sigma x = \underline{59}$

$\Sigma x^2 = \underline{203}$

b.

x	f	xf	$x^2 f$
1	3	3	3
2	4	8	16
3	6	18	54
4	5	20	80
5	2	10	50
	20	59	203

$\Sigma xf = 59$

$\Sigma x^2 f = 203$

c. (1) The sum of four 2's is 8, the sum of six 3's is 18, and so on, as can be seen on both charts.
(2) The sum of four $(2)^2$'s is 16, the sum of six $(3)^2$'s is 54, and so on, as can be seen on both charts.
(3) $\Sigma x = 59$ and so does Σxf.
(4) $\Sigma x^2 = 203$ and so does $\Sigma x^2 f$.

2. a.

x	f	xf	$x^2 f$
61	2	122	7442
64	4	256	16384
67	7	469	31423
70	9	630	44100
73	3	219	15987
	25	1696	115336

$\Sigma f = 25$

$\Sigma xf = 1696$

$\Sigma x^2 f = 115{,}336$

b. $\bar{x} = \dfrac{\Sigma xf}{\Sigma f} = \dfrac{1696}{25} = 67.84 = \underline{67.8}$

$s = \sqrt{\dfrac{\Sigma x^2 f - \dfrac{(\Sigma xf)^2}{\Sigma f}}{(\Sigma f - 1)}} = \sqrt{\dfrac{115{,}336 - [(1696)(1696)/25]}{(24)}}$

$s = \sqrt{11.64} = 3.41 = \underline{3.4}$

SOLUTIONS TO EXERCISES 2-6

1. a.

x	f	xf	$x^2 f$
5	2	10	50
10	1	10	100
15	5	75	1,125
20	13	260	5,200
25	10	250	6,250
30	5	150	4,500
35	2	70	2,450
40	1	40	1,600
45	1	45	2,025
Sum	40	910	23,300

SOLUTIONS: Chapter 2

$$\bar{x} = \frac{910}{40} = \underline{22.75}$$

$$s = \sqrt{\frac{(23,300) - [(910)(910)/40]}{(39)}} = \sqrt{66.60256} = 8.161 = \underline{8.16}$$

b. $\bar{x} \pm 2s = 22.75 \pm 2(8.16)$

22.75 − 16.32 to 22.75 + 16.32

6.43 to 39.07

The proportion between 6.43 and 39.07 is 36/40 or $\underline{0.90}$.

SOLUTIONS TO SELF-CORRECTING EXERCISES FOR CHAPTER 2

1. a. $n = 4$ b. $\Sigma x = 24$ c. $\bar{x} = 24/4 = 6.0$
2. a. $n = 6$ b. $\Sigma x = 21$ c. $\bar{x} = 21/6 = 3.5$
3. a. The response variable x is the number of hours spent to complete the exam by each student.
   b. $n = 8$ c. $\Sigma x = 72$ d. $\bar{x} = 72/8 = 9.0$
4. a. The response variable x is the number of hours spent studying during the weekend by each student.
   b. Continuous; values reported are whole hours, but time is a measurable quantity.
   c. $n = 12$ d. $\Sigma x = 84$ e. $\bar{x} = 84/12 = 7.0$
5. $\bar{x} = 6.0$

x	$x - \bar{x}$	$(x - \bar{x})^2$
6	0	0
4	−2	4
5	−1	1
9	3	9
24	0 ⓒⓚ	14

$s^2 = 14/3 = 4.67 = \underline{4.7}$

6. $\bar{x} = 3.5$

x	$x - \bar{x}$	$(x - \bar{x})^2$
3	-0.5	0.25
5	1.5	2.25
1	-2.5	6.25
3	-0.5	0.25
2	-1.5	2.25
7	3.5	12.25
21	0 ck	23.50

$s^2 = 23.50/5 = \underline{4.7}$

7. a. $\bar{x} = 9.0$

x	$x - \bar{x}$	$(x - \bar{x})^2$
9	0	0
6	-3	9
8	-1	1
9	0	0
11	2	4
7	-2	4
9	0	0
13	4	16
72	0 ck	34

$s^2 = 34/7 = 4.857 = \underline{4.9}$

b.

x	x^2
9	81
6	36
8	64
9	81
11	121
7	49
9	81
13	169
72	682

$$s^2 = \frac{682 - \frac{(72)(72)}{8}}{(7)} = 4.857 = \underline{4.9}$$

SOLUTIONS: Chapter 2

c. $s = \sqrt{4.857} = 2.2022 = \underline{2.2}$

8. a. $\bar{x} = 7.0$

x	$x - \bar{x}$	$(x - \bar{x})^2$
4	-3	9
8	1	1
12	5	25
3	-4	16
2	-5	25
7	0	0
6	-1	1
9	2	4
9	2	4
13	6	36
6	-1	1
5	-2	4
84	0 ck	126

$s^2 = 126/11 - 11.4545 - \underline{11.5}$

b.

x	4	8	12	3	2	7	6	9	9	13	6	5	84
x^2	16	64	144	9	4	49	36	81	81	169	36	25	714

$$s^2 = \frac{714 - \frac{(84)(84)}{12}}{(11)} = 11.4545 = \underline{11.5}$$

c. $s = \sqrt{11.4545} = 3.384 = \underline{3.4}$

9. a. The response variable x is the regular price of an item on sale with a $2 rebate.
   b. continuous (an amount of money)
   c. $\bar{x} = \Sigma x/n = 102/7 = 14.571 = \underline{14.6}$

SOLUTIONS: Chapter 2

d.

x	x^2
10	100
13	169
16	256
17	289
15	225
20	400
11	121
102	1560

$$s = \sqrt{\frac{1560 - \frac{(102)(102)}{7}}{(6)}}$$

$$= \sqrt{12.2857} = 3.5050 = \underline{3.5}$$

10. a.

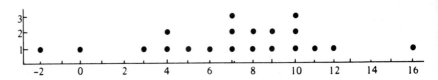

b. Estimates will vary.

c. $n = 20$, $\Sigma x = 144$, $\Sigma x^2 = 1364$

$$\bar{x} = \frac{144}{20} = \underline{7.2}$$

$$s = \sqrt{\frac{1364 - \frac{(144)(144)}{20}}{(19)}} = \sqrt{17.2210}$$

$$= 4.1498 = \underline{4.1}$$

11. $\Sigma x = 816$

a. $\bar{x} = \frac{\Sigma x}{n} = \frac{816}{8} = \underline{102.0}$

b. Since $\bar{x}$ is a whole number and there are only eight pieces of data, the easiest method for calculating the standard deviation is by using formula (2-7).

SOLUTIONS: Chapter 2

x	$x - \bar{x}$	$(x - \bar{x})^2$
100	-2	4
100	-2	4
101	-1	1
101	-1	1
102	0	0
103	1	1
104	2	4
105	3	9
816	0 (ck)	24

$$s = \sqrt{\frac{\Sigma(x - \bar{x})^2}{n - 1}}$$

$$= \sqrt{24/7}$$

$$= \sqrt{3.43} = \underline{1.85}$$

12. $u = x - 100$, code by subtracting 100.

x	x^2
100.0	10,000.00
100.4	10,080.16
101.0	10,201.00
101.4	10,281.96
101.8	10,363.24
103.2	10,650.24
104.4	10,899.36
105.2	11,067.04
817.4	83,543.00

a. $\bar{x} = \Sigma x/n = 817.4/8 = 102.175 = \underline{102.2}$

b. $s_x = \sqrt{\dfrac{\Sigma x^2 - \dfrac{(\Sigma x)^2}{n}}{(n-1)}}$

$= \sqrt{\dfrac{83543.00 - \dfrac{(817.4)^2}{8}}{(7)}} = \sqrt{\dfrac{25.155}{7}}$

$= \sqrt{3.594} = 1.896 = \underline{1.90}$

c. Rounding off the data had very little effect on the calculated results. In a situation where the numbers are of similar magnitude, the effect of rounding off would be comparable.

13. a. A frequency histogram is a bar graph showing the frequency and the class marks of each class.

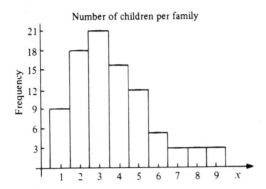

Notice that class marks (x) are labeled along the horizontal axis of the histogram.

b. A relative frequency histogram has a vertical scale of relative frequency; that is, frequency/total number. Instead of showing nine families with one child, it shows 9/90 (9 out of the 90 in the sample).

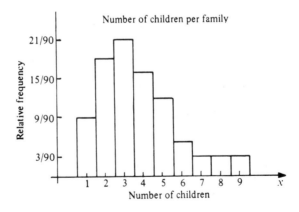

Since each class has only one value, the class mark (that value) is labeled along the horizontal axis.

SOLUTIONS: Chapter 2

14. a. Recall that "ungrouped" means all the same values of x are collected together.

x	f
0	3
1	6
2	4
3	4
4	10
5	2
6	4
7	5
8	6
9	6
	50

b. A stem-and-leaf diagram shows exactly the same information as an ungrouped frequency distribution, namely, the frequency of each number value.

```
0 | 0  0  0
1 | 1  1  1  1  1  1
2 | 2  2  2  2  2
3 | 3  3  3  3  3
4 | 4  4  4  4  4  4  4  4  4  4
5 | 5  5
6 | 6  6  6  6
7 | 7  7  7  7  7
8 | 8  8  8  8  8  8
9 | 9  9  9  9  9  9
```

c. A frequency histogram presents the same information as parts a and b in a bar-graph form.

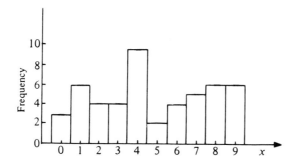

SOLUTIONS: Chapter 2

d. A cumulative frequency distribution shows a list of values for x (classes) and a corresponding list of cumulative frequencies. That is, the frequency shown opposite each class is the total number of pieces of data in that class and in all smaller valued classes. Notice that the cumulative frequency for the last class (largest valued class) is equal to the n of the sample.

x	Cum. freq.
0	3
1	9
2	13
3	17
4	27
5	29
6	33
7	38
8	44
9	50

e. An ogive is a graphic (line graph) representation of the cumulative frequency distribution. The vertical scale shows relative frequencies.

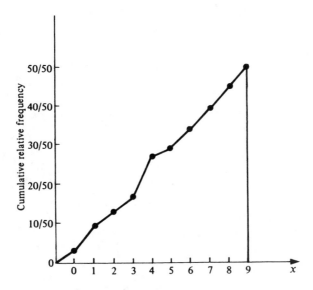

SOLUTIONS: Chapter 2

15. a. A grouped frequency distribution differs from the ungrouped in that each class has more than one value. Class 1 in this list (9–13) contains all the 9's, 10's, 11's, 12's, and 13's.

Class limits	Frequency	Class mark
9–13	3	11
14–18	15	16
19–23	26	21
24–28	23	26
29–33	8	31
34–38	2	36
39–43	3	41

b. A relative frequency histogram identifies the vertical scale that is to be used on the graphic presentation.

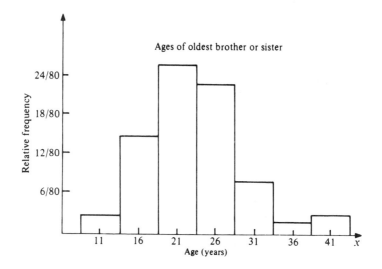

c. A cumulative grouped frequency distribution requires a list showing two columns: class identification and cumulative frequencies.

Class limits	Cum. freq.
9–13	3
14–18	18
19–23	44
24–28	67
29–33	75
34–38	77
39–43	80

d.

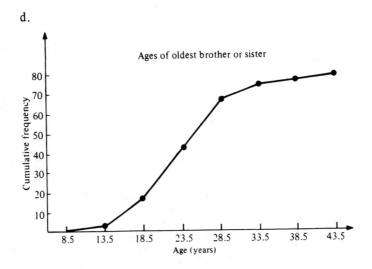

SOLUTIONS: Chapter 2

16. a. The list of 50 weights, when fit into the specified class limits, forms the following frequency distribution.

Class limits	Frequency
28.0–30.9	1
31.0–33.9	1
34.0–36.9	2
37.0–39.9	12
40.0–42.9	10
43.0–45.9	9
46.0–48.9	8
49.0–51.9	7
	$\Sigma f = 50$

b.

[Histogram: Weight of luggage; x-axis "Weight (pounds)" with boundaries 27.95, 30.95, 33.95, 36.95, 39.95, 42.95, 45.95, 48.95, 51.95; y-axis Frequency with marks at 4, 8, 12, 16.]

c. To construct an ogive, you will first need to form a cumulative frequency distribution. Using the answer to part a, the cumulative frequencies are:

Class limits	Cum. freq.
28.0–30.9	1
31.0–33.9	2
34.0–36.9	4
37.0–39.9	16
40.0–42.9	26
43.0–45.9	35
46.0–48.9	43
49.0–51.9	50

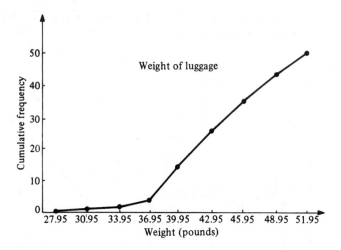

Note: Both of these graphic presentations use class boundaries to label the horizontal axis. Each class contains many decimal values; thus the boundary values are the best choice for the histogram. The ogive requires the class boundary.

17. a. Each class mark can be found by adding the respective class limits and dividing by two. First class mark: $(21 + 25) \div 2 = 23$, and so forth.

Class limits	Class mark
21–25	23
26–30	28
31–35	33
36–40	38
41–45	43
46–50	48
51–55	53
56–60	58
61–65	63

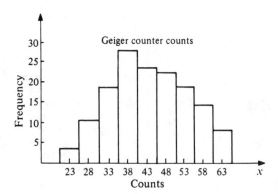

SOLUTIONS: Chapter 2

b. The upper class boundaries for each class and the cumulative frequencies are graphed to form an ogive.

Class boundaries	Cum. freq.
20.5–25.5	3
25.5–30.5	14
30.5–35.5	33
35.5–40.5	61
40.5–45.5	85
45.5–50.5	108
50.5–55.5	127
55.5–60.5	142
60.5–65.5	150

18. a. "Ungrouped frequency distribution" means that only the like number values are collected together. The frequency number is the number of times each value of the data occurred.

x	2	3	5	6	7	8	9	10	11	12	13	14	15	16	17	19
f	1	2	1	2	3	2	5	4	4	6	2	2	3	1	1	1

b.
```
 2 | 2
 3 | 3  3
 4 |
 5 | 5
 6 | 6  6
 7 | 7  7  7
 8 | 8  8
 9 | 9  9  9  9  9
10 | 0  0  0  0
11 | 1  1  1  1
12 | 2  2  2  2  2  2
13 | 3  3
14 | 4  4
15 | 5  5  5
16 | 6
17 | 7
18 |
19 | 9
```

c.

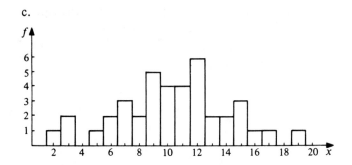

d. The best way to find these summations is to use a chart showing the extensions xf and x^2f for each value of x, then find the column totals. (A calculator is helpful here.)

x	f	xf	x^2f
2	1	2	4
3	2	6	18
5	1	5	25
6	2	12	72
7	3	21	147
8	2	16	128
9	5	45	405
10	4	40	400
11	4	44	484
12	6	72	864
13	2	26	338
14	2	28	392
15	3	45	675
16	1	16	256
17	1	17	289
19	1	19	361
	40	414	4858

e. Formula (2-2) is the correct formula to use here, since the set of data is in the form of a frequency distribution.

$$\bar{x} = \Sigma xf/\Sigma f = 414/40 = \underline{10.35}$$

SOLUTIONS: Chapter 2

f. The standard deviation is found by taking the square root of the variance, which is calculated by using formula (2-11).

$$s^2 = \frac{\Sigma x^2 f - \frac{(\Sigma xf)^2}{n}}{(n-1)} ; n = \Sigma f$$

$$= \frac{4858 - \frac{(414)(414)}{40}}{(39)}$$

$$= 14.695$$

$$s = \sqrt{14.695} = \underline{3.83}$$

One of the purposes of using a grouped frequency distribution is to simplify the calculations and reduce the number of classifications. Therefore, the histogram presents a better image of the distribution. Both of these ideas are demonstrated by the next four answers.

g. The classes become 2-4, 5-7, 8-10, and so on.

Class limits	f
2-4	3
5-7	6
8-10	11
11-13	12
14-16	6
17-19	2
	40

h.

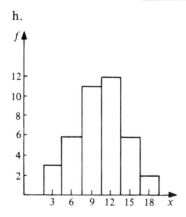

i.

Class mark x	f	xf	x^2f
3	3	9	27
6	6	36	216
9	11	99	891
12	12	144	1728
15	6	90	1350
18	2	36	648
	40	414	4860

Notice that the two summations found in part i are very similar to the summations found earlier in part d.

j. Formula (2-2) is used, and the mean is:

$$\bar{x} = \Sigma xf/n = 414/40 = \underline{10.35}$$

k. Formula (2-11) is used to find the variance; then the square root is taken to find the standard deviation.

$$s^2 = \frac{\Sigma x^2 f - \frac{(\Sigma xf)^2}{n}}{(n-1)} = \frac{4860 - \frac{(414)(414)}{40}}{(39)} = 14.746$$

$$s = \sqrt{14.746} = \underline{3.84}$$

l.

	from d, e, f	from i, j, k
Σf	40	40
Σxf	414	414
$\Sigma x^2 f$	4858	4860
$\bar{x}$	10.35	10.35
s	3.83	3.84

The differences are caused by the approximation that took place in the grouping. For example, in the first class in part i, the grouping shows three 3's, when actually it was one 2 and two 3's.

19. a. A chart showing the extensions xf and $x^2 f$ is used with the frequency distribution from Exercise 13.

SOLUTIONS: Chapter 2

x	f	xf	x^2f
1	9	9	9
2	18	36	72
3	21	63	189
4	16	64	256
5	12	60	300
6	5	30	180
7	3	21	147
8	3	24	192
9	3	27	243
	90	334	1588

$\Sigma xf = 334$

$\Sigma x^2 f = 1588$

b. Formula (2-2) is used to calculate the mean of a set of data represented by a frequency distribution where the extensions xf and $x^2 f$ are used.

$$\bar{x} = \Sigma xf/n = 334/90 = 3.71$$

c. Formula (2-11) is used to find the variance of a frequency distribution. The square root of variance gives the standard deviation.

$$s^2 = \frac{\Sigma x^2 f - (\Sigma xf)^2}{(n-1)} = \frac{1588 - \frac{(334)(334)}{90}}{(89)}$$

$$= 3.916$$

$$s = \sqrt{3.92} = 1.98$$

20. a. A chart showing the extensions xf and $x^2 f$ is used with the frequency distribution from Exercise 14.

x	f	xf	x^2f
0	3	0	0
1	6	6	6
2	4	8	16
3	4	12	36
4	10	40	160
5	2	10	50
6	4	24	144
7	5	35	245
8	6	48	384
9	6	54	486
	50	237	1527

$\Sigma xf = 237$

$\Sigma x^2 f = 1527$

SOLUTIONS: Chapter 2

b. Formula (2-2) is used to calculate the mean of a set of data represented by a frequency distribution where the extensions xf and x^2f are used.

$\bar{x} = \Sigma xf/n = 237/50 = \underline{4.74}$

c. Formula (2-11) is used to find the variance of a frequency distribution. The square root of variance gives the standard deviation.

$$s^2 = \frac{\Sigma x^2 f - (\Sigma xf)^2}{(n-1)} = \frac{1527 - \frac{(237)(237)}{50}}{(49)}$$

$= 8.237$

$s = \sqrt{8.24} = \underline{2.87}$

21. a. Using the class marks and the frequencies found in Exercise 15, we need to construct a chart showing the extensions xf and x^2f.

x	f	xf	x^2f
11	3	33	363
16	15	240	3840
21	26	546	11466
26	23	598	15548
31	8	248	7688
36	2	72	2592
41	3	123	5043
	80	1860	46540

$\Sigma xf = \underline{1860}$

$\Sigma x^2 f = \underline{46540}$

b. Formula (2-2) is used to calculate the mean of a grouped frequency distribution. The only difference between the ungrouped and the grouped frequency distributions is that the class mark in an ungrouped distribution has a value equal to all data belonging to that class, while the grouped distribution has several values assigned to each class. Once these values belong to the class, they are treated as though their value were that of the class mark.

$\bar{x} = \Sigma xf/n = 1860/80 = \underline{23.25}$

SOLUTIONS: Chapter 2

c. $s^2 = \dfrac{\Sigma x^2 f - \dfrac{(\Sigma xf)^2}{n}}{(n-1)} = \dfrac{46540 - \dfrac{(1860)(1860)}{80}}{(79)}$

$= 41.71$

$s = \sqrt{41.71} = \underline{6.46}$

22.

Class limits	Class mark, x	f	xf	$x^2 f$
28.0–30.9	29.45	1	29.45	867.3025
31.0–33.9	32.45	1	32.45	1,053.0025
34.0–36.9	35.45	2	70.90	2,513.4050
37.0–39.9	38.45	12	461.40	17,740.8300
40.0–42.9	41.45	10	414.50	17,181.0250
43.0–45.9	44.45	9	400.05	17,782.2225
46.0–48.9	47.45	8	379.60	18,012.0200
49.0–51.9	50.45	7	353.15	17,816.4175
		50	2,141.50	92,966.2250

The above summations are to be used with formulas (2-2) and (2-11) to calculate the mean and standard deviation.

$\bar{x} = \dfrac{\Sigma xf}{n}$

$\bar{x} = \dfrac{2141.50}{50} = \underline{42.83}$

$s^2 = \dfrac{\Sigma x^2 f - \dfrac{(\Sigma xf)^2}{n}}{(n-1)} = \dfrac{1245.78}{(49)}$

$s^2 = 25.424$

$s = \sqrt{25.424}$

$s = 5.042 = \underline{5.04}$

SOLUTIONS: Chapter 2

23.

Class mark, x	f	xf	x^2f
23	3	69	1,587
28	11	308	8,624
33	19	627	20,691
38	28	1,064	40,432
43	24	1,032	44,376
48	23	1,104	52,992
53	19	1,007	53,371
58	15	870	50,460
63	8	504	31,752
	150	6,585	304,285

Use formulas (2-2) and (2-11).

$$\bar{x} = \frac{\Sigma xf}{n}$$

$$\bar{x} = \frac{6585}{150} = \underline{43.9}$$

$$s^2 = \frac{\Sigma x^2 f - (\Sigma xf)^2}{(n-1)} = \frac{304{,}285 - \frac{(6585)^2}{150}}{(149)}$$

$$= 102.0369$$

$$s = \sqrt{102.0369} = \underline{10.10}$$

24. a. The median's position is $i = (n+1)/2 = (50+1)/2 = 25.5$. The median is therefore halfway between the 25th and the 26th pieces of data.

$$\tilde{x} = \frac{1+1}{2} = \underline{1}$$

b. The mode is 0, the value with the greatest frequency.
c. $nk/100 = (50)(25)/100 = 12.5$; therefore, $i = $ 13th; $Q_1 = 0$
d. $nk/100 = (50)(30)/100 = 15.0$; therefore, $i = $ 15.5th; $P_{30} = 0$.
e. The range is the difference between the largest-valued and the smallest-valued data.

$$\text{range} = 7 - 0 = \underline{7}$$

SOLUTIONS: Chapter 2

25. The weights should be placed in rank order (smallest to largest) to facilitate finding the measures requested.

		Weights in rank order		
1	28.9	38.9	41.5	44.4	47.6
2	33.7	39.1	41.6	45.3	47.8
3	34.5	39.7	41.8 –P_{45}	45.6	48.5
4	35.1	39.8	42.5	45.7	49.2
5	37.1	39.9	42.5 ⎫ $\tilde{x}$	45.8	49.2
6	37.3	39.9	42.6 ⎭	46.2	49.7
7	38.3	40.2	43.1	46.8	50.1
8	38.4	40.3	43.7	46.8 –Q_3	50.1
9	38.5	40.7	43.9	47.2	50.9
10	38.6	40.8	44.3	47.6	51.4

a. Median's position = $i = (n + 1)/2 = (50 + 1)/2 = 25.5$

Median = $(42.5 + 42.6)/2 = \underline{42.55}$

(Halfway between the 25th and 26th piece of data counting in from either end.)

b. Q_3's position: $nk/100 = (50)(75)/100 = 37.5; i = 38; Q_3 = \underline{46.8}$

c. P_{45}'s position: $nk/100 = (50)(45)/100 = 22.5; i = 23; P_{45} = \underline{41.8}$

26. To find the median and the quartiles we need to arrange the data in rank order. Rank order can be shown graphically on a dot-array diagram; therefore, let us construct such a graph in place of the ranked list.

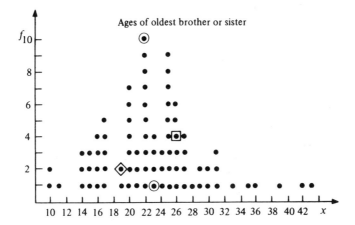

a. The median's position = $i = (n+1)/2 = (80+1)/2 = 40.5$. The 40th and 41st values are circled on the dot-array diagram above. (Count from the lower valued end.) The median is midway between the two values 22 and 23; therefore, $\tilde{x} = \underline{22.5}$.
b. Midquartile = $(Q_1 + Q_3)/2$
$Q_1: nk/100 = (80)(25)/100 = 20.0; i = 20.5$

$$Q_1 = \frac{19 + 20}{2} = \underline{19.5}$$

(Q_1 is halfway between the two values identified by the diamonds.)
$Q_3: i = 20.5$ from the largest-valued data

$$Q_3 = \frac{26 + 26}{2} = \underline{26}$$

(Q_3 is halfway between the two values identified by the squares.)

$$\text{midquartile} = \frac{19.5 + 26}{2} = \underline{22.75}$$

c. H represents the value of the largest valued piece of data (43) and L represents the value of the smallest valued piece of data (10).

$$\text{midrange} = \frac{H + L}{2} = \frac{43 + 10}{2} = \underline{26.5}$$

d. Range = $H - L = 43 - 10 = \underline{33}$

CHAPTER 3

SOLUTIONS TO EXERCISES 3-1

1. a. As children get older, they grow taller. Therefore r would be <u>positive</u>. (Both variables increase.)
   b. The higher the rank, the higher the pay. <u>Positive</u>.
   c. The more intelligent are expected to solve problems faster. (More intelligent, less time.) <u>Negative</u>.
   d. One expects to spend more for repairs on an older car. (Both variables increase in value.) <u>Positive</u>.
   e. Older cars tend to be worth less as trade-ins. (Age increases, value decreases.) <u>Negative</u>.
   f. The students with higher midterm averages tend to get the higher final grades in any course. <u>Positive</u>.

SOLUTIONS TO EXERCISES 3-2

1. a. The summation, $\Sigma(y - \hat{y})^2$, must be as small as possible.
   b. The least that $\Sigma(y - \hat{y})^2$ could ever be is zero, since each $(y - \hat{y})^2$ must be nonnegative. This would occur only if each (x,y) of the data were to lie exactly on the line of best fit.

SOLUTIONS: Chapter 3

2. When x is the input and y is the output variable, the least-squares criterion is to make $\Sigma(y - \hat{y})^2$ as small as possible. By interchanging the roles of x and y, the least-squares criterion would call for the $\Sigma(x - \hat{x})^2$ to be as small as possible, where $\hat{x} = c_0 + c_1 y$ (c_0 and c_1 being the replacements for the b_0 and b_1 values). Making $\Sigma(y - \hat{y})^2$ as small as possible would in no way guarantee that $\Sigma(x - \hat{x})^2$ was at a minimum value.

3. A slope of zero would mean that the line of best fit was in a horizontal position and that a change in x had no effect on the resulting y-value. A negative slope would mean that the value of y is decreasing as x increases in value. A positive slope indicates that y is increasing in value as x increases.

SOLUTIONS TO EXERCISES 3-3

1. The main use of the line of best fit is to aid in predicting the value of y that will occur.

2. a. To predict a value for y is to calculate the $\hat{y}$ that corresponds to a given x-value.
   b. The $\hat{y}$-value that results is expected to be the average value of y that corresponds to the x-value used to determine $\hat{y}$.

3. $\hat{y} = 17.5$, when $x = 3$ $[\hat{y} = 10 + 2.5(3)]$
   $\hat{y} = 30.0$, when $x = 8$ $[\hat{y} = 10 + 2.5(8)]$

SOLUTIONS TO SELF-CORRECTING EXERCISES FOR CHAPTER 3

1. a. not linear
   b. Not linear; perhaps it's parabolic.
   c. linear
   d. not linear

   Notice that the points in the scatter diagram for part c tend to lie along a straight band whereas the other three suggest a path that is not a straight line.

SOLUTIONS: Chapter 3

2. a.

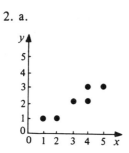

b.

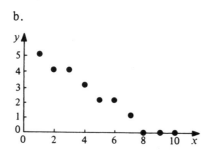

c.

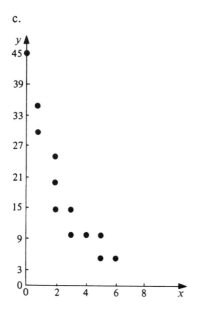

Note: Each pair of values from the chart is represented by a point (x,y) on the xy coordinate-axis system.

SOLUTIONS: Chapter 3

3. a. *D*; points lie in an up-hill position and the relationship appears to be relatively strong, perhaps +0.8.
   b. *B*; points lie in a down-hill position and appear to be of relatively moderate strength, perhaps −0.5.
   c. *C*; the value of *y* does not appear to be affected by a change in *x*; therefore, *r* is approximately equal to 0.
   d. *A*; down-hill position and all points lie exactly on a straight line; therefore, $r = -1$.
   e. *B*; down-hill position and all points seem to lie close to the line; thus *r* might be −0.8.
   f. *D*; similar to *B* except that the points lie in an up-hill position. Perhaps $r = +0.5$.

4. a. Positive; as a boy grows older, his feet grow longer.
   b. Negative; as a ball ages it tends to deaden.
   c. Near zero; age of a person and age of his auto seem completely unrelated in pattern.
   d. Positive; generally, a person's salary increases as his length of service increases.
   e. Near zero; it seems unlikely that commuting distance would have any effect on a student's grade-point average.

Note: (1) *r* is positive when *y* increases in size as *x* increases.
(2) *r* is negative when *y* decreases in size as *x* increases.

5. a.

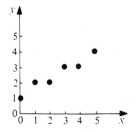

b.

x	y	x^2	xy	y^2
0	1	0	0	1
1	2	1	2	4
2	2	4	4	4
3	3	9	9	9
4	3	16	12	9
5	4	25	20	16
15	15	55	47	43

SOLUTIONS: Chapter 3

c. $SS(xy) = (47) - \dfrac{(15)(15)}{6} = 9.5$

$SS(x) = 55 - \dfrac{(15)^2}{6} = 17.5$

$SS(y) = 43 - \dfrac{(15)^2}{6} = 5.5$

$r = \dfrac{9.5}{\sqrt{(17.5)(5.5)}} = 0.968 = \underline{0.97}$

d. $b_1 = \dfrac{9.5}{17.5}$

$b_1 = 0.54286$

$b_1 = \underline{0.54}$

$b_0 = \dfrac{1}{6}[15 - (0.54286)(15)]$

$b_0 = \dfrac{1}{6}[6.8571]$

$b_0 = 1.1428 = \underline{1.14}$

b_0 is the y-intercept, the value of y at the point where the line of best fit crosses the y-axis. If you inspect the graph in part a, the y-intercept appears to be a value near +1.

b_1 is the slope of the line of best fit. This value is the rate at which the value of y changes as x increases by one unit. The graph suggests that the value of y increases by about 2 as x increases by 4. Thus the slope, b_1, is approximately 1/2.

e. $\hat{y} = 1.14 + 0.54x$

Recall that the equation of a line is the mathematical relationship between the x- and y-coordinates of every point on that line.

6. a.

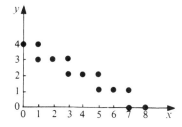

b.

x	y	x^2	xy	y^2
0	4	0	0	16
1	4	1	4	16
1	3	1	3	9
2	3	4	6	9
3	3	9	9	9
3	2	9	6	4
4	2	16	8	4
5	2	25	10	4
5	1	25	5	1
6	1	36	6	1
7	1	49	7	1
7	0	49	0	0
8	0	64	0	0
52	26	288	64	74

c. $SS(xy) = 64 - \left[\dfrac{(52)(26)}{13}\right] = -40.0$

$SS(x) = 288 - \dfrac{(52)^2}{13} = 80.0$

$SS(y) = 74 - \dfrac{(26)^2}{13} = 22.0$

$r = \dfrac{-40.0}{\sqrt{(80.0)(22.0)}} = -0.95346 = \underline{-0.95}$

d. $b_1 = \dfrac{-40.0}{80.0} = \underline{-0.50}$

$b_0 = \dfrac{1}{13}[26 - (-0.50)(52)] = \underline{4.00}$

The scatter diagram suggests that the y-intercept is 4 while the slope is a negative 1/2. As x increases, the y-value decreases. Compare this information to the information and comment in Exercise 5.

e. $\hat{y} = 4.0 + 0.5x$ (See comment in Exercise 5e.)

SOLUTIONS: Chapter 3

7. a.

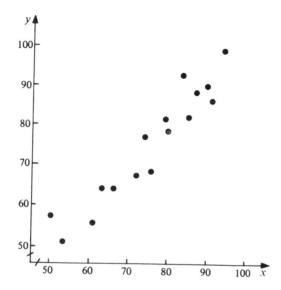

b.

x	y	x^2	xy	y^2
76	68	5776	5168	4624
61	55	3721	3355	3025
74	77	5476	5698	5929
94	99	8836	9306	9801
79	81	6241	6399	6561
63	64	3969	4032	4096
80	78	6400	6240	6084
87	88	7569	7656	7744
91	86	8281	7826	7396
50	57	2500	2850	3249
83	93	6889	7719	8649
85	82	7225	6970	6724
66	64	4356	4224	4096
72	67	5184	4824	4489
53	50	2809	2650	2500
90	89	8100	8010	7921
1204	1198	93332	92927	92888

c. $SS(xy) = 92,927 - \dfrac{(1,204)(1,198)}{16} = 2,777.5$

$SS(x) = 93,332 - \dfrac{(1,204)^2}{16} = 2,731.0$

$SS(y) = 92,888 - \dfrac{(1,198)^2}{16} = 3,187.75$

$r = \dfrac{2777.5}{\sqrt{(2,731.0)(3,187.75)}} = 0.9413 = \underline{0.94}$

d. $b_1 = \dfrac{2,777.5}{2,731.0} = 1.01703 = \underline{1.02}$

$b_0 = \dfrac{1}{16}[1198 - (1.01703)(1,204)] = -1.6565 = \underline{-1.66}$

$\hat{y} = -1.66 + 1.02x$

(See comments in Exercises 5 and 6.)

Note that the y-intercept is not shown on the graph. The vertical scale shown on the graph is actually located at $x = 45$, not at $x = 0$; thus the value of b_0 is actually -1.66, not approximately 40 as the graph might lead you to think.

8. a.

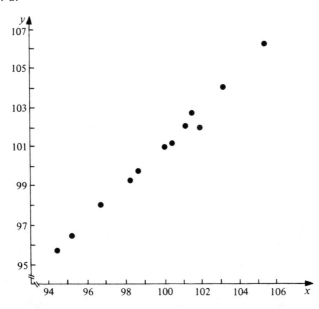

b. $n = 12$, $\Sigma x = 1196.0$, $\Sigma y = 1208.0$, $\Sigma x^2 = 119312.24$, $\Sigma xy = 120504.00$, $\Sigma y^2 = 121708.64$.

Only the totals have been shown here. These totals are obtained by use of a table showing the pairs of x- and y-values with the extensions and totals as seen in Exercise 7.

c. $SS(xy) = 120{,}504.00 - \dfrac{(1{,}196.0)(1{,}208.0)}{12}$

$ = 106.66667$

$SS(x) = 119{,}312.24 - \dfrac{(1{,}196.0)^2}{12} = 110.90667$

$SS(y) = 121{,}708.64 - \dfrac{(1{,}208.0)^2}{12} = 103.30667$

$r = \dfrac{106.66667}{\sqrt{(110.906667)(103.30667)}} = 0.9965 = \underline{0.997}$

d. $b_1 = \dfrac{106.66667}{110.90667} = 0.96$

$b_0 = \dfrac{1}{12}[1208.0 - (0.96177)(1196.0)] = 4.81$

$\hat{y} = \underline{4.81 + 0.96x}$

e. The evidence shown here certainly seems to agree with the statement. (See comments in Exercises 5 through 7.) In order to determine the expected rectal temperature for a patient with a given oral temperature, you replace the x with the given oral temperature. The resulting value for $\hat{y}$ is the expected rectal temperature.

9. $\hat{y} = 4.81 + 0.96x$
   a. $x = 100.0$, $\hat{y} = 4.81 + 0.96(100.0) = \underline{100.81}$
   b. $x = 98.0$, $\hat{y} = 4.81 + 0.96(98.0) = \underline{98.89}$
   c. $x = 102.0$, $\hat{y} = 4.81 + 0.96(102.0) = \underline{102.73}$
   d. $x = 98.6$, $\hat{y} = 4.81 + 0.96(98.6) = \underline{99.47}$
   e. $x = 103.4$, $\hat{y} = 4.81 + 0.96(103.4) = \underline{104.07}$

   f.

x	$\hat{y}$	$x + 1$
100.0	100.8	101.0
98.0	98.9	99.0
102.0	102.7	103.0
98.6	99.5	99.6
103.4	104.1	104.4

   $\hat{y}$, the calculated value, using the line of best fit of the data, appears to be approximately two-tenths of a degree below $x + 1$, the temperature that is to be expected according to the claim.

10. $\hat{y} = -1.66 + 1.02x$
   a. $x = 60, \hat{y} = -1.66 + 1.02(60) = \underline{59.54}$
   b. $x = 80, \hat{y} = -1.66 + 1.02(80) = \underline{79.94}$
   c. $x = 92, \hat{y} = -1.66 + 1.02(92) = \underline{92.18}$
   (See comments in Exercise 9.)

CHAPTER 4

SOLUTIONS TO EXERCISES 4-1

1. a. true   b. true

2. a. Yes; their solution sets have no elements in common.
   b. No; the entire sample space is not covered. Zero has not been accounted for.
   c. No; the occurrence of 2 means that both events have occurred.
   d. Yes; the entire sample space has been accounted for, that is, each element of the sample space belongs to one or more events.
   e. event $D = \{0\}$
   f. event $E = \{4, 5, 6, 7, 8, 9\}$

SOLUTIONS TO EXERCISES 4-2

1. a. Sample Space = H1, H2, H3, H4, H5, H6, T1, T2, T3, T4, T5, T6
   b. $A \cap B$ = H2, H4, H6
   c. $P(A \text{ and } B) = n(A \cap B)/n(S) = 3/12 = \underline{1/4}$
   d. $A \cup B = \{$H1, H2, H3, H4, H5, H6, T2, T4, T6$\}$
           a   ab   a   ab   a   ab   b   b   b

   Note: The a's and b's under the set elements indicate why they belong to the set. The a means that this element belongs to A, the b means that it belongs to B, and ab means that it belongs to both. Recall that the union contains all those elements which belong to either one of the sets or both of the sets.

   e. $P(A \text{ or } B) = n(A \cup B)/n(s) = 9/12 = \underline{3/4}$

2. a. $P(W \text{ even and } B \text{ even}) = n(W \text{ even and } B \text{ even})/n(S) = 9/36 = \underline{1/4}$
   b. $P(W \text{ even or } B \text{ even}) = n(W \text{ even or } B \text{ even})/n(S) = 27/36 = \underline{3/4}$
   c. $P(5 \text{ or } B \text{ odd}) = 20/36 = \underline{5/9}$
   d. $P(5 \text{ and } B \text{ odd}) = 2/36 = \underline{1/18}$
   e. $P(11 \text{ or } B \text{ is } 3) = 8/36 = \underline{2/9}$
   f. $P(11 \text{ and } B \text{ is } 3) = 0/36 = \underline{0}$

SOLUTIONS TO EXERCISES 4-3

1. a. $P(A, \text{knowing } B \text{ did not happen}) = 3/6 = \underline{1/2}$
   b. $P(B, \text{knowing } A) = 3/6 = \underline{1/2}$
   c. $P(B, \text{knowing } \bar{A}) = 3/6 = \underline{1/2}$
2. a. $P(S) = 6/36 = \underline{1/6}$ and $P(S, \text{knowing } T) = 0/3 = \underline{0}$
   b. No, they are not independent. The probability of event S was influenced by knowledge of event T.
   c. $P(S) = \underline{1/6}$ and $P(S, \text{knowing } F) = \underline{1/6}$
   d. Yes, S and F are independent. The probability of event S did not change due to knowledge of event F.
   e. $P(F) = 6/36 = \underline{1/6}$ and $P(F, \text{knowing } \bar{T}) = \underline{5/33}$
   f. No, F and T are not independent. $P(F)$ and $P(F, \text{knowing } \bar{T})$ are not equal.
   g. $P(F) = 6/36 = \underline{1/6}$ and $P(F, \text{knowing } S) = \underline{1/6}$
   h. Yes, F and S are independent. Yes, these results were found in parts c and d.
3. $P(H)$, the probability of a head on any given toss, is $1/2$.
   In experiment I, $P(H \text{ on 1st toss}) = 1/2$, as is $P(H \text{ on 2nd toss, knowing } H \text{ on 1st})$, and $P(H \text{ on 2nd toss, knowing } T \text{ on 1st})$, and so on. Therefore the trials are independent.
   In experiment II, $P(H \text{ on 1st toss}) = 1/2$. However, $P(H \text{ on 2nd toss, knowing } H \text{ on 1st}) = 0$, as the trial will not take place. Thus we would conclude that the result on the first trial does influence the second trial. Therefore, they are dependent trials.
4. a. $12/52$ b. $4/52$ c. $40/52$ d. $4/12$ e. $0/40$
   f. F and K are dependent events. $P(K) \neq P(K, \text{knowing } F) \neq P(K, \text{knowing } \bar{F})$.

SOLUTIONS TO EXERCISES 4-4

1. a. $P(R \text{ and } S) = P(R) \cdot P(S) = (1/4)(1/2) = \underline{1/8}$
   b.

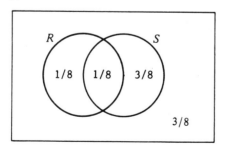

$P(S) = 1/2; P(R|S) = P(R \text{ and } S) \div P(S) = (1/8) \div (4/8) = 1/4$; and $P(R)$ is given to be $1/4$, so $P(R) = P(R|S)$.

244 SOLUTIONS: Chapter 4

2. $P(A) \cdot P(B) \neq P(A \text{ and } B)$
$(1/3)(1/4) = 1/12 \neq 1/5$
Therefore, A and B are dependent events.
$P(A) = 20/60 = 1/3; P(B) = 15/60 = 1/4$.
$P(A|B) = (12/60) \div (15/60) = 12/15 = 4/5$
$1/3 \neq 4/5$, therefore A and B are dependent.

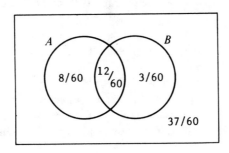

3. $P(F \text{ and } K) = n(F \text{ and } K)/n(S) = \underline{4/52}$
or
$P(F \text{ and } K) = P(F) \cdot P(K, \text{knowing } F) = (12/52)(4/12) = \underline{4/52}$

SOLUTIONS TO EXERCISES 4-5

1. $P(A \text{ or } B) = P(A) + P(B) - P(A \text{ and } B)$
$= 0.3 + 0.4 - (0.3)(0.4)$, since A and B are independent.
$P(A \text{ or } B) = \underline{0.58}$

2. $P(C \text{ or } D) = P(C) + P(D)$, since they are mutually exclusive.
$P(C \text{ or } D) = 0.75 + 0.125 = \underline{0.875}$

3. a. $P(F \text{ or } K) = P(F) + P(K) - P(F \text{ and } K) = 12/52 + 4/52 - 4/52 = \underline{12/52}$
b. $P(F \text{ or } \bar{K}) = P(F) + P(\bar{K}) - P(F \text{ and } \bar{K}) = 12/52 + 48/52 - 8/52 = 52/52$
$= \underline{1}$
c. $P(\bar{F} \text{ or } K) = P(\bar{F}) + P(K) - P(\bar{F} \text{ and } K) = 40/52 + 4/52 - 0/52 = \underline{44/52}$ (K and $\bar{F}$ are mutually exclusive events.)

SOLUTIONS TO SELF-CORRECTING EXERCISES FOR CHAPTER 4

1. AB   BC   CD   DE   (10 different pairs of choices)
AC   BD   CE
AD   BE
AE

Note: AB and BA would represent the same pair of choices, thus only one of them is listed. The same is true for all other pairs.

2. a. 1/10; 1 out of 10 chances.
   b. 5/10 or 1/2; 5 are even out of the 10.
   c. 8/10; the set "less than 8" contains 0, 1, 2, 3, 4, 5, 6, 7; a total of 8 out of the 10 numbers.
   d. 4/10; the set "less than 8 and odd" contains 1, 3, 5, 7; a total of 4 out of the set of 10 numbers.
   e. Separately, "odd" is 1, 3, 5, 7, 9; and "less than 5" is 0, 1, 2, 3, 4. The solution set for "either odd or less than 5" contains 0, 1, 2, 3, 4, 5, 7, 9. Therefore, 8 out of the 10 are either odd or less than 5. Answer: 8/10.
3. 4/52 or 1/13; there are four jacks in a deck of 52 cards. Each should have an equal chance of being selected.
4. a. 5/26 (5 of the 26 letters are vowels.)
   b. 6/26 (6 out of the 26 letters.)
5. 30-day months = (April, June, September, November)
   P(30-day month) = 4/12 or 1/3

6. Refer to the table on page 130 in the textbook. The sample space contains 36 equally like sample points.
   P(event) = n(event)/n(sample space)
   a. 11 = {(5,6), (6,5)}: P(11) = 2/36
   b. 7 or 11 = {(1,6), (2,5), (3,4), (4,3), (5,2), (6,1), (5,6), (6,5)}: P(7 or 11) = 8/36
   c. P(not 6) = 31/36   (1 − 5/36)
   "Not 6" is the complement of "6." Thus the total of their probabilities is exactly 1. n(6) = 5 since 6 = {(1,5), (2,4), (3,3), (4,2), (5,1)}.
   d. P(more than 8) = P(9) + P(10) + P(11) + P(12) = 10/36
   "More than 8" is composed of 4 mutually exclusive events; thus we need only to add their individual probabilities.
   e. P(total less than 18) = 1   (all totals are less than 18)

7. a. P(2 children) = 6/50 = 0.12   [P(2) = n(2)/50]
   b. P(4 or more) = P(4) + P(5) + P(6) + P(7) + P(9) = 20/50 = 0.40
   (See comment in Exercise 6d.)
   c. True; no family could possibly belong to an "intersection" of these sets.

8. a. Sample space (probability):

   RR(1/9)   RW(1/9)   RB(1/9)   Note: Each of these nine
   WR(1/9)   WW(1/9)   WB(1/9)   events is equally
   BR(1/9)   BW(1/9)   BB(1/9)   likely.

   Note: WR represents white on first drawing and red on the second.
   b. P(both white) = P(WW) = 1/9
   c. P(both blue) = P(BB) = 1/9
   d. P(same color) = P(RR or WW or BB) = 3/9
   e. P(neither is white) = P(RR or RB or BR or BB) = 4/9

246   SOLUTIONS: Chapter 4

f. P(at least one is blue) = P(RB or WB or RB or BW or BR) = **5/9**
or
P(at least one is blue) = 1 − P(no blue) = 1 − P(RR,RW,WR,WW) = 1 − 4/9 = **5/9**

Each of the five probabilities above is found by dividing the number of elements in the solution set for each event by 9, the number of elements in the sample space shown above.

Note: Random selection implies that each person had an equal chance of being selected.

9. a. P(both are blind) = P(1st is blind) · P(2nd is blind, knowing the 1st was blind) = (5/25)(4/24) = 1/30. The 4/24 represents the 4 blind of the 24 who are still unselected.
   b. P(neither are blind) = P(1st is not blind) · P(2nd is not blind, knowing the 1st was not blind) = (20/25)(19/24) = **19/30**

10. a. 15/20 = 3/4
    b. P(all 3 are good) = P(1st is good) · P(2nd is good, if the 1st was good) · P(3rd is good, if the 1st and 2nd were good)
    P(all 3 are good) = (15/20)(14/19)(13/18) = **91/228**

11. a. "Equally likely sample space" representation:

	RW_1	RW_2	RB_1	RB_2	RB_3
$W_1 R$		$W_1 W_2$	$W_1 B_1$	$W_1 B_2$	$W_1 B_3$
$W_2 R$	$W_2 W_1$		$W_2 B_1$	$W_2 B_2$	$W_2 B_3$
$B_1 R$	$B_1 W_1$	$B_1 W_2$		$B_1 B_2$	$B_1 B_3$
$B_2 R$	$B_2 W_1$	$B_2 W_2$	$B_2 B_1$		$B_2 B_3$
$B_3 R$	$B_3 W_1$	$B_3 W_2$	$B_3 B_1$	$B_3 B_2$	

Sample space (probability):

$$S = \left\{ RW\left(\frac{4}{30}\right), RB\left(\frac{6}{30}\right), WW\left(\frac{2}{30}\right), WB\left(\frac{12}{30}\right), BB\left(\frac{6}{30}\right) \right\}$$

b. P(both red) = 0
c. P(both white) = 2/30
d. P(both blue) = 6/30
e. P(1st drawn is red) = 5/30
f. P(2nd drawn is red) = 5/30
g. P(neither is red) = 20/30

SOLUTIONS: Chapter 4

12. a. Sample space:

```
T  | .  .  .  •
O  | .  •  •  .
O  | .  •  •  .
T  | .  .  .  •
   |_____
     B  O  O  T
```

Probability for each point is $\frac{1}{16}$.

b. $P(\text{same letter}) = P(\text{both are T's or both are O's}) = 6/16 = \underline{3/8}$

or

$P(\text{same}) = P(\text{both T's}) + P(\text{both O's})$ [mutually exclusive]
$\quad = P(\text{T from TOOT}) \cdot P(\text{T from BOOT})$
$\quad\quad + P(\text{O from TOOT}) \cdot P(\text{O from BOOT})$
$\quad = (1/2)(1/4) + (1/2)(1/2)$

$P(\text{same}) = 1/8 + 1/4 = \underline{3/8}$

13. a. $P(\text{both }\$10) = (2/7)(1/6) = \underline{1/21}$
    b. $P(\text{both }\$1) = (5/7)(4/6) = \underline{10/21}$
    c. $P(\text{one }\$1\text{ and one }\$10) = P(\$1\text{ first, then }\$10) + P(\$10\text{ first, then }\$1)$
    $\quad = (5/7)(2/6) + (2/7)(5/6)$
    $\quad = \underline{10/21}$

    or

    $1 - [P(\text{both }\$10) + P(\text{both }\$1)] = 1 - [(1/21) + (10/21)] = \underline{10/21}$

    or — consider the sample space idea.

		$\$1_1$	$\$1_2$	$\$1_3$	$\$1_4$	$\$1_5$	$\$10_1$	$\$10_2$
First draw	$\$1_1$		x	x	x	x	+	+
	$\$1_2$	x		x	x	x	+	+
	$\$1_3$	x	x		x	x	+	+
	$\$1_4$	x	x	x		x	+	+
	$\$1_5$	x	x	x	x		+	+
	$\$10_1$	+	+	+	+	+		0
	$\$10_2$	+	+	+	+	+	0	

Second draw

X represents both $1 $n(\text{X}) = 20$

O represents both $10 $n(\text{O}) = 2$

+ represents one of each $n(+) = 20$

$P(\text{both }\$1) = 20/42;\ P(\text{both }\$10) = 2/42;\ P(\text{one of each}) = 20/42$

14. a.

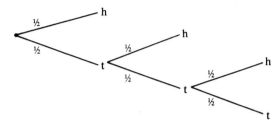

Sample space (probability):

$S = \{h(1/2), th(1/4), tth(1/8), ttt(1/8)\}$

b. H = (h, th, tth)
T = (ttt)
H and T are mutually exclusive events.

c. H and T are not independent. If H occurs, there is no way that T can happen; therefore, knowledge that H has happened causes the probability of T to become zero.

15. Tree diagram representation:

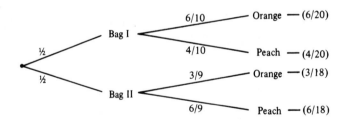

a. P(an orange is selected) = P(bag I and orange) + P(bag II and orange)
$= (1/2)(6/10) + (1/2)(3/9) = 6/20 + 3/18$
$= 0.300 + 0.16\overline{6} = \underline{0.467}$
b. P(a peach is selected) = P(bag I and peach) + P(bag II and peach)
$= (1/2)(4/10) + (1/2)(6/9)$
$= 0.200 + 0.33\overline{3} = \underline{0.53\overline{3}}$
or
P(peach) = $1 - P$(orange) = $1 - 0.467 = 0.53\overline{3}$
c. P(orange from bag II) = $3/18$
d. P(orange, knowing bag I) = $6/10$
e. P(bag II, knowing an orange was selected)

SOLUTIONS: Chapter 4

$$= \frac{P(\text{orange from bag II})}{P(\text{orange})} = \frac{3/18}{(6/20) + (3/18)}$$

$$= \frac{0.16\overline{6}}{0.46\overline{7}} = \underline{0.357}$$

Note: This is the relative proportion of oranges that come from bag II.

16. a. $P(\text{seeds}) = P(\text{pink with seeds}) + P(\text{white with seeds})$
        $= 0.30 + 0.40 = \underline{0.70}$
    b. $P(\text{seedless}) = 1.0 - P(\text{seeds}) = 1.0 - 0.70 = \underline{0.30}$
    c. $P(\text{white}) = P(\text{white seedless}) + P(\text{white with seeds})$
        $= 0.10 + 0.40 = \underline{0.50}$
    d. $P(\text{pink}) = 1 - P(\text{white}) = 1.0 - 0.50 = \underline{0.50}$
    e. $P(\text{pink and seedless}) = P(\text{pink seedless}) = \underline{0.20}$
    f. $P(\text{pink or with seeds}) = P(\text{pink}) + P(\text{seeds}) - P(\text{pink with seeds})$
        $= 0.50 + 0.70 - 0.30 = \underline{0.90}$
    g. $P(\text{pink, knowing it has seeds})$ is the proportion of those with seeds that are also pink.

    $$P(\text{pink, knowing it has seeds}) = \frac{0.30}{0.70} = \underline{0.43}$$

    h. $P(\text{seeds, knowing it is pink})$ is the proportion of the pink that have seeds.

    $$P(\text{seeds, knowing it is pink}) = \frac{0.30}{0.50} = \underline{0.60}$$

17. a. $P(S) = 4/16 = 1/4 = 0.25$
    b. $P(F) = 7/16 = 0.4375$
    c. $P(T) = 6/16 = 0.375$
    d. $P(S \text{ and } F) = P(4,4) = 1/16 = 0.0625$
    e. $P(F \text{ and } T) = 0/16 = 0.0$
    f. $P(S \text{ and } T) = P((1,1) \text{ or } (2,2)) = 2/16 = 0.125$
    g. $P(S \text{ or } F) = 10/16 = 0.625$
    h. $P(F \text{ or } T) = 13/16 = 0.8125$
    i. $P(S \text{ or } T) = 8/16 = 0.50$
    j. The sample space with the events identified:

	1	2	3	4
1	ST	T	T	F
2	T	ST	•	F
3	T	•	S	F
4	F	F	F	SF

S and F intersect at (4,4); therefore, they are not mutually exclusive.
S and T intersect at (1,1) and (2,2); therefore, they are not mutually exclusive.
F and T do not intersect; therefore, they are mutually exclusive.

18. a. $P(A \text{ and } B) = P(A) \cdot P(B)$ since A and B are independent.

    $P(A \text{ and } B) = (1/2)(2/5) = 1/5 = \underline{0.20}$

    b. $P(\text{not } B) = 1.0 - P(B) = 1.0 - (2/5) = 3/5 = \underline{0.60}$
    c. $P(A \text{ or } B) = P(A) + P(B) - P(A \text{ and } B)$, since A and B are not given as mutually exclusive.

    $P(A \text{ or } B) = (1/2) + (2/5) - (1/5) = \underline{0.70}$

19. a. $P(A \text{ or } B) = P(A) + P(B)$, since A and B are mutually exclusive.

    $P(A \text{ or } B) = 0.5 + 0.3 = \underline{0.8}$

    b. $P(A \text{ and } B) = \underline{0.0}$; since A and B are mutually exclusive, they cannot both occur at the same time.

20. a. $P(C \text{ or } D) = P(C) + P(D) - P(C \text{ and } D)$
    $= P(C) + P(D) - P(C) \cdot P(D)$
    $= 0.7 + 0.2 - (0.7)(0.2)$
    $= 0.90 - 0.14 = \underline{0.76}$

    b. $P(C \text{ and } D) = P(C) \cdot P(D)$
    $= (0.7)(0.2) = \underline{0.14}$

21. $P(\text{lung}) = 0.3, P(\text{blood}) = 0.4, P(\text{both}) = 0.1$
    a. $P(\text{at least one}) = P(\text{lung or blood})$
    $= P(\text{lung}) + P(\text{blood}) - P(\text{both})$
    $= 0.3 + 0.4 - 0.1 = \underline{0.6}$
    b. If they are independent, then $P(\text{lung and blood})$ will equal the product of $P(\text{lung})$ and $P(\text{blood})$.

    $P(\text{both}) = P(\text{lung and blood}) = P(\text{lung}) \cdot P(\text{blood})$, if independent.

    $P(\text{both}) = 0.10$ while $P(\text{lung}) \cdot P(\text{blood}) = (0.3)(0.4) = 0.12$

    Since these values are different, we must conclude that the two events are *not* independent.

22. $P(A) = 0.4, P(B) = 0.2$
    a. $P(A \text{ and } B) = P(A) \cdot P(B)$, since independent.
    $P(A \text{ and } B) = (0.4)(0.2) = \underline{0.08}$

b. $P(\text{neither}) = P(\bar{A} \text{ and } \bar{B}) = P(\bar{A}) \cdot P(\bar{B})$, since independent.
$P(\bar{A} \text{ and } \bar{B}) = [1 - P(A)] \cdot [1 - P(B)] = (0.6)(0.8) = \underline{0.48}$
c. $P(\text{at least one is success}) = 1.0 - P(\text{neither})$
$= 1.0 - 0.48 = \underline{0.52}$

23. $P(\text{more ore}) = P[(\text{buy A and ore}) \text{ or } (\text{buy B and ore})]$
$= P(\text{buy A})P(\text{ore on A}) + P(\text{buy B})P(\text{ore on B})$
$- P[(\text{buy A and ore on A}) \text{ and } (\text{buy B and ore on B})]$
$= (0.5)(0.2) + (0.6)(0.1) - (0.5)(0.2)(0.6)(0.1)$
$= 0.100 + 0.060 - 0.006 = \underline{0.154}$

24. $P(\text{A's product does well}) = P(\text{competitor does not market and}$
A's product does well)
$+ P(\text{competitor does market and}$
A's product does well)
$= (0.4)(0.9) + (0.6)(0.5)$
$= 0.36 + 0.30 = \underline{0.66}$

CHAPTER 5

SOLUTIONS TO EXERCISES 5-1

1. a. $\bar{x}$: mean of a sample
   s: standard deviation of a sample
   μ: mean of a population
   σ: standard deviation of a population
   b. $\bar{x}$ and μ are both means. $\bar{x}$ is the mean value of the data belonging to a sample. μ is the mean value of all the data possible from the population.
   c. s and σ are both measures of the standard deviation found among the individual values of the variable. s is the measure associated with the sample, while σ is the corresponding measure of dispersion for the entire population.

SOLUTIONS: Chapter 5

2. a.

x	$P(x) = \dfrac{5-x}{10}$	$x \cdot P(x)$	$x^2 \cdot P(x)$
1	4/10	4/10	4/10
2	3/10	6/10	12/10
3	2/10	6/10	18/10
4	1/10	4/10	16/10
	10/10 = 1.0	20/10 = 2.0	50/10 = 5.0

$\mu = \Sigma x \cdot P(x) = \underline{2.0}$

$\sigma = \sqrt{\Sigma x^2 \cdot P(x) - [\Sigma x \cdot P(x)]^2} = \sqrt{5.0 - (2.0)^2} = \sqrt{1.0} = \underline{1.0}$

b. The distribution in this exercise is simply the reverse of the distribution used as an illustration. Thus the standard deviation has the same value since the "spread" is unchanged.

3. a.

x	$P(x)$	$xP(x)$	$x^2 P(x)$
1	1/9	1/9	1/9
2	2/9	4/9	8/9
4	3/9	12/9	48/9
5	2/9	10/9	50/9
9	1/9	9/9	81/9
	9/9 = 1.0	36/9 = 4.0	188/9 = 20.88

$\mu = \underline{4.0}$

$\sigma = \sqrt{20.88 - (4.0)^2} = \sqrt{4.88} = \underline{2.21}$

SOLUTIONS TO EXERCISES 5-2

1. a. Experiment: inspecting each case.
   Trial: inspecting each item, 12 trials per experiment.
   Outcome: defective, not defective.
   $p = 0.01$, $x =$ number of defectives in one case.
   b. $n = 12, x = 0, p = 0.01; P(x = 0) = \underline{0.886}$

SOLUTIONS: Chapter 5

c. $P(\text{no more than one defective}) = P(x = 0,1) = P(x = 0) + P(x = 1)$
$= 0.886 + 0.107 = \underline{0.993}$

2. Experiment: taking quiz by guessing.
Trial: answering of one question by guessing.
Outcome: correct answer or incorrect answer.
x = number of correct answers obtained on quiz.
a. $n = 10, p = 0.2, x = 5; p(x = 5) = \underline{0.026}$
b. $P(x < 5) = P(x = 0,1,2,3,4)$
$= P(x = 0) + P(x = 1) + P(x = 2) + P(x = 3) + P(x = 4)$
$= 0.107 + 0.268 + 0.302 + 0.201 + 0.088 = \underline{0.966}$
c. $P(x = 2) = 0.302$

d.

1,2	1,3	1,4	1,5	1,6	1,7	1,8	1,9	1,10		9
	2,3	2,4	2,5	2,6	2,7	2,8	2,9	2,10		8
		3,4	3,5	3,6	3,7	3,8	3,9	3,10		7
			4,5	4,6	4,7	4,8	4,9	4,10		6
				5,6	5,7	5,8	5,9	5,10		5
					6,7	6,8	6,9	6,10		4
						7,8	7,9	7,10		3
							8,9	8,10		2
								9,10		+1
										45

e. $P(1,2) = \dfrac{1}{5} \cdot \dfrac{1}{5} \cdot \dfrac{4}{5} \cdot \dfrac{4}{5} \cdot \dfrac{4}{5} \cdot \dfrac{4}{5} \cdot \dfrac{4}{5} \cdot \dfrac{4}{5} \cdot \dfrac{4}{5} \cdot \dfrac{4}{5}$

$= \left(\dfrac{1}{5}\right)^2 \cdot \left(\dfrac{4}{5}\right)^8$

f. $P(x = 2) = (45)\left[\left(\dfrac{1}{5}\right)^2 \cdot \left(\dfrac{4}{5}\right)^8\right]$

SOLUTIONS TO SELF-CORRECTING EXERCISES FOR CHAPTER 5

1. A probability distribution is a list that pairs the various values of the random variable x (sum of the two rankings in this case) with the appropriate probability.

x	P(x)
2	1/16
3	2/16
4	3/16
5	4/16
6	3/16
7	2/16
8	1/16
	16/16 = 1 (ck)

2. The mean and variance of a probability distribution are found by using formulas (5-2) and (5-4). The standard deviation is then found by taking the square root of the variance.

x	P(x)	xP(x)	$x^2P(x)$
1	.1	.1	.1
3	.2	.6	1.8
5	.4	2.0	10.0
7	.2	1.4	9.8
9	.1	.9	8.1
	1.0	5.0	29.8

$\mu = \Sigma x P(x)$

$\mu = 5.0$

$\sigma = \sqrt{\Sigma x^2 P(x) - [\Sigma x P(x)]^2}$

$\sigma = \sqrt{29.8 - (5.0)^2}$

$= \sqrt{4.8}$

$= 2.19$

3. See the comments in Exercise 2.

x	P(x)	xP(x)	$x^2P(x)$
1	.1	.1	.1
2	.2	.4	.8
3	.3	.9	2.7
4	.2	.8	3.2
5	.2	1.0	5.0
	1.0	3.2	11.8

$\mu = \Sigma x P(x) = 3.2$

$\sigma = \sqrt{\Sigma x^2 \cdot P(x) - [\Sigma x P(x)]^2}$

$\sigma = \sqrt{11.8 - (3.2)^2} = \sqrt{1.56}$

$= 1.25$

4. The solution of this problem is similar to Exercises 2 and 3. See the comments in Exercise 2.

x	$P(x)$	$xP(x)$	$x^2P(x)$
3	.1	.3	.9
4	.2	.8	3.2
5	.3	1.5	7.5
6	.4	2.4	14.4
	1.0	5.0	26.00

a. $\mu = \Sigma xP(x) = \underline{5.0}$

b. $\sigma = \sqrt{\Sigma x^2 P(x) - [\Sigma xP(x)]^2}$
   $= \sqrt{26.0 - (5.0)^2}$
   $= \sqrt{1.0} = \underline{1.0}$

5. Each of the questions involves binomial events. Each toss of a coin represents a trial in the particular experiment being carried out. Thus n equals the number of coins being tossed. In each question, x is equal to n; therefore, the probability formula on page 186 becomes

$$P(x = n) = \binom{n}{n} \cdot p^n \cdot q^0 = p^n \quad \text{or} \quad \underline{(1/2)^n}$$

a. $P(\text{all heads}) = P(x = 2)$, when $n = 2, p = 1/2$.

$P(x = 2) = (1/2)^2 = \underline{1/4}$

b. $P(\text{all heads}) = P(x = 3)$, when $n = 3, p = 1/2$.

$P(x = 3) = (1/2)^3 = \underline{1/8}$

c. $P(\text{all heads}) = P(x = 4)$, when $n = 4, p = 1/2$.

$P(x = 4) = (1/2)^4 = \underline{1/16}$

d. $P(\text{all heads}) = P(x = 10)$, when $n = 10, p = 1/2$.

$$P(x = 10) = \left(\frac{1}{2}\right)^{10} = \underline{\frac{1}{1024}}$$

6. a. This inspection is actually a binomial experiment with each walkie-talkie representing a trial. Five inspected means that $n = 5$. Finding all to work properly means that we are only interested in x, the number found to work, equal to 5.
$p = 0.90$. ($n = 5, p = 0.90, x = 5, q = 0.10$.) Note: p is the probability a walkie-talkie works.
$P(\text{accepting shipment}) = P(x = 5)$, when $n = 5, p = 0.9$.

$P(x = 5) = (0.9)^5 = \underline{0.590}$

b. The situation is the same as above except that $p = 0.80$.
$P(\text{accepting shipment}) = P(x = 5)$, when $n = 5, p = 0.8$.

$P(x = 5) = (0.8)^5 = \underline{0.328}$

7. The tossing of three identical coins is a binomial experiment. Each coin becomes a trial with two possible outcomes. Therefore, $n = 3$, $p = 0.5$, and x is the number of heads observed.
   a. $P(x = 3) = 0.125$
   b. $P(x = 2) = 0.375$
   c. $P(x = 1) = 0.375$
   d. $P(x = 0) = 0.125$
   e. $P(x = 1,2,3) = 0.875$
   f. $P(x = 0,3) = 0.25$

8. The selecting of two male students and observing vet or non-vet is a binomial experiment with each selected student representing a trial. $n = 2$, $p = 0.1$, and x is the number of vets. (Use Table 4 in textbook.)
   a. $P(\text{both are vets}) = P(x = 2) = \underline{0.01}$
   b. $P(\text{neither is a vet}) = P(x = 0) = \underline{0.81}$
   c. $P(\text{exactly one is a vet}) = P(x = 1) = \underline{0.18}$

9. This is not a true binomial probability problem; however, it seems reasonable to use the binomial distribution to approximate the probabilities for the events below. $n = 8$ (the eight shots), $p = 0.8$ (probability of a bull's-eye), and x is the number of bull's-eyes that occur in the eight shots. (Use Table 4 on pages 520-522 in the textbook.)
   a. $P(\text{8 hits in 8 tries}) = P(x = 8) = \underline{0.168}$
   b. $P(\text{at least 7 hits}) = P(x = 7) + P(x = 8) = .336 + .168 = \underline{0.504}$
   c. $P(\text{at least 6 hits}) = P(x = 6) + P(x = 7) + P(x = 8)$
      $= .294 + .336 + .168 = \underline{0.798}$

10. Assume independence; $n = 6$, $p = .5$, $x = n(\text{boys})$.
    a. $P(x = 6) = \underline{0.016}$
    b. $P(\text{youngest is a girl and other 5 are boys}) = P(\text{1st is boy and 2nd is boy and 3rd is boy and } \ldots \text{ and 6th is girl}) = (0.5)^6 = \underline{0.016}$
    c. $P(\text{1 girl and 5 boys}) = P(x = 5) = \underline{0.094}$
    d. $P(\text{2 girls and 4 boys}) = P(x = 4) = \underline{0.234}$
    e. $P(\text{3 girls and 3 boys}) = P(x = 3) = \underline{0.312}$
    f. i. $n = 6$
       ii. $P(\text{boy}) = 0.5$
       iii. two outcomes: boy or girl
       iv. we assumed independence of trials

11. a. $P(\text{tails}) = 3 [P(\text{heads})]$; therefore, $P(\text{tails}) = 3/4$ and $P(\text{heads}) = 1/4$.
    $P(\text{2 heads occur in 2 flips}) = P(x = 2)$, when $n = 2$ and $p = 1/4$.
    $$P(x = 2) = (1/4)^2 = 1/16 = \underline{0.0625}$$

    b. The expected number of heads would be the mean value for the distribution of x, the number of heads.
    $$P(x = 0) = \binom{2}{0}\left(\frac{1}{4}\right)^0 \left(\frac{3}{4}\right)^2 = \frac{9}{16}$$

SOLUTIONS: Chapter 5

$$P(x=1) = \binom{2}{1}\left(\frac{1}{4}\right)^1\left(\frac{3}{4}\right)^1 = \frac{6}{16}$$

$$P(x=2) = \binom{2}{2}\left(\frac{1}{4}\right)^2\left(\frac{3}{4}\right)^0 = \frac{1}{16}$$

x	$P(x)$	$x \cdot P(x)$
0	9/16	0/16
1	6/16	6/16
2	1/16	2/16
	16/16	8/16

$\mu = 8/16 = 1/2$; therefore, the expected number of heads to be seen on two flips of this coin is **1/2**.

12. Similar to Exercise 9, this is not a true binomial probability experiment. However, the binomial distribution should give a reasonable estimate for each of the following probabilities. Each patient in a set represents a trial ($n = 4$) and $p = 0.6$, while x is the number of recoveries found per set of 4 patients.
   a. $P(x = 3) = \underline{0.346}$
   b. $\mu = np = (4)(0.6) = \underline{2.4}$
   c. $\sigma = \sqrt{npq} = \sqrt{(2.4)(0.4)} = \sqrt{0.96} = \underline{0.98}$

13. $P(\text{tail}) = 2\,P(\text{head})$; so $P(\text{tails}) = 2/3$, $P(\text{heads}) = 1/3$.
    Binomial experiment; $n = 3$, $p = 1/3$, $x = n(\text{heads})$.
    a.

x	$P(x)$
0	$\binom{3}{0}\left(\frac{1}{3}\right)^0\left(\frac{2}{3}\right)^3 = \frac{8}{27}$
1	$\binom{3}{1}\left(\frac{1}{3}\right)^1\left(\frac{2}{3}\right)^2 = \frac{12}{27}$
2	$\binom{3}{2}\left(\frac{1}{3}\right)^2\left(\frac{2}{3}\right)^1 = \frac{6}{27}$
3	$\binom{3}{3}\left(\frac{1}{3}\right)^3\left(\frac{2}{3}\right)^0 = \frac{1}{27}$
	$\frac{27}{27}$

b.

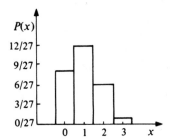

c.

x	$P(x)$	$xP(x)$	$x^2P(x)$
0	8/27	0/27	0/27
1	12/27	12/27	12/27
2	6/27	12/27	24/27
3	1/27	3/27	9/27
	27/27	27/27	45/27

$\mu = \Sigma xP(x)$ $\sigma = \sqrt{\Sigma x^2 P(x) - [\Sigma xP(x)]^2}$
$= 27/27 = \underline{1.0}$ $= \sqrt{45/27 - (27/27)^2} = \sqrt{1.667 - 1.0}$
                               $= \sqrt{0.6667}$
                               $= \underline{0.82}$

d. n repeated trials: three tosses
two outcomes: heads or tails
independence: each toss is separate and results have no effect on later tosses.
$p = P(\text{heads}) = 1/3, q = P(\text{tails}) = 2/3$
$x = n(\text{heads})$ that occur in a set of three tosses

e. $\mu = np = (3)(1/3) = \underline{1.0}$
$\sigma = \sqrt{npq} = \sqrt{3(1/3)(2/3)} = \sqrt{2/3} = \sqrt{0.667} = \underline{0.82}$
[same results as found in part c]

14. a. $P(6) = 2P(1); P(1) = P(2) = P(3) = P(4) = P(5)$; therefore, $P(6) = 2/7, P(1) = 1/7, P(2) = 1/7$, and so forth.

SOLUTIONS: Chapter 6

b.

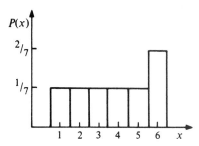

c.

x	$P(x)$	$xP(x)$	$x^2P(x)$
1	1/7	1/7	1/7
2	1/7	2/7	4/7
3	1/7	3/7	9/7
4	1/7	4/7	16/7
5	1/7	5/7	25/7
6	2/7	12/7	72/7
	7/7 = 1.0	27/7	127/7

$\mu = \Sigma xP(x) = 27/7 = \underline{3.86}$

$\sigma = \sqrt{\Sigma x^2 P(x) - [\Sigma xP(x)]^2}$

$= \sqrt{127/7 - (27/7)^2}$

$= \sqrt{18.14 - 14.88}$

$= \sqrt{3.26} = \underline{1.81}$

d. Not binomial; there are six outcomes to be concerned with, not just two.

CHAPTER 6

SOLUTIONS TO EXERCISES 6-1

1. $z = \dfrac{x - \mu}{\sigma}$, $\mu = 27.8$, and $\sigma = 4.2$.

    a. $z = \dfrac{27.8 - 27.8}{4.2} = \underline{0.0}$ b. $z = \dfrac{32.0 - 27.8}{4.2} = \underline{1.0}$

    c. $z = \dfrac{35.2 - 27.8}{4.2} = \underline{1.76}$ d. $z = \dfrac{25.4 - 27.8}{4.2} = \underline{-0.57}$

e. $z = \dfrac{12.1 - 27.8}{4.2} = \underline{-3.74}$ f. $z = \dfrac{29.4 - 27.8}{4.2} = \underline{0.38}$

2. a. False; only 99.7% of the data will be between $z = -3$ and $z = +3$ when the distribution is normal. For other distributions this percentage will vary.
   b. False; the standard score can be used with any distribution that has a mean and a standard deviation.

3. $z = \dfrac{x - \mu}{\sigma}$; so $x = \mu + z \cdot \sigma$; $\mu = 125.9$;

   a. $x = 125.9 + (1.2)(8.7) = \underline{136.34}$
   b. $x = 125.9 + (1.5)(8.7) = \underline{138.95}$
   c. $x = 125.9 + (0.5)(8.7) = \underline{130.25}$
   d. $x = 125.9 + (-2.4)(8.7) = \underline{105.02}$
   e. $x = 125.9 + (4.25)(8.7) = \underline{162.875}$
   f. $x = 125.9 + (-3.75)(8.7) = \underline{93.275}$

SOLUTIONS TO EXERCISES 6-2

1. a. 0.2704   b. 0.3554   c. 0.4842
   d. 0.4699   e. 0.4994   f. 0.49997

2. a. $0.5000 - 0.3554 = \underline{0.1446}$ b. $0.5000 + 0.3554 = \underline{0.8554}$
   c. $0.5000 + 0.4842 = \underline{0.9842}$ d. $0.5000 - 0.4842 = \underline{0.0158}$
   e. $0.3554 + 0.49997 = \underline{0.85537}$ f. $0.49997 - 0.3554 = \underline{0.14457}$

SOLUTIONS TO SELF-CORRECTING EXERCISES FOR CHAPTER 6

1. These values are obtained directly from Table 5.
   a. 0.3413   b. 0.2190   c. 0.4535   d. 0.4994

2. These values are obtained directly from Table 5, using the idea of symmetry. (See Illustration 6-3, page 208 of textbook.)
   a. 0.3907   b. 0.2190   c. 0.4995   d. 0.4826

3. These values are obtained directly from Table 5.
   a. 0.4850   b. 0.4625   c. 0.1628   d. 0.2967

4. These values are obtained directly from Table 5.
   a. 0.4998   b. 0.4826

Note: The three concepts, area under the normal curve, probability of an individual value, and proportion of the population, are synonymous and are all treated alike.

SOLUTIONS: Chapter 6

5. To obtain these values, we need to find two values from Table 5 and then add or subtract, whichever is appropriate.
   a. 0.3413 + 0.4938 = 0.8351
   b. 0.4370 + 0.4693 + 0.9063
   c. 0.4332 + 0.4332 = 0.8664
   d. 0.4896 + 0.4991 = 0.9887
   e. 0.4993 − 0.3413 = 0.1580

6. See the note above and the comment in Exercise 5.
   a. 0.4332 + 0.4990 = 0.9322
   b. 0.4990 − 0.4332 = 0.0658
   c. 0.4990 − 0.4332 = 0.0658
   d. 0.4990 + 0.4332 = 0.9322

7. See the note above and the comment in Exercise 5.
   a. 0.3849 + 0.4918 = 0.8767
   b. 0.4918 − 0.3849 = 0.1069

8. These values are obtained by adding or subtracting the table value with 0.5000, the area under half of the curve.
   a. 0.5000
   b. 0.5000 − 0.3413 = 0.1587
   c. 0.5000 − 0.4893 = 0.0107
   d. 0.5000 + 0.3023 = 0.8023
   e. 0.5000 + 0.4099 = 0.9099
   f. 0.5000 − 0.4821 = 0.0179

9. See the comment in Exercise 8.
   a. 0.5000 + 0.4948 = 0.9948
   b. 0.5000 − 0.4948 = 0.0052
   c. 0.5000 + 0.4357 = 0.9357
   d. 0.5000 − 0.4938 = 0.0062
   e. 0.5000 − 0.3413 = 0.1587

10. See the comment in Exercise 8.
    a. 0.5000
    b. 0.5000 − 0.2190 = 0.2810
    c. 0.5000 + 0.2190 = 0.7190

11. This exercise is a mixture of the several exercises above.
    a. 0.3289
    b. 0.3944
    c. 0.4332 + 0.4332 = 0.8664
    d. 0.4878 + 0.4994 = 0.9872
    e. 0.5000 + 0.4994 = 0.9994
    f. 0.5000 + 0.4918 = 0.9918

12. The answer to each of these questions is obtained by using Table 5 (in the back of the textbook) conversely from its usage in the preceding eleven exercises. The probability is given and must be converted to an area between $z = 0$ and some positive z_0. Then the probability is found (in Table 5) and the answer read from the right side and top of the table.

    a.
    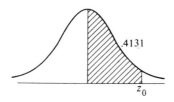
    $z_0 = 1.36$

b.

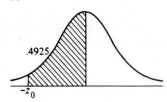

$z_0 = 2.43$

$-z_0 = \underline{2.43}$

c.

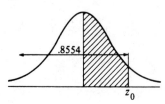

$0.8554 - 0.5000 = 0.3554$

m (shaded area) = 0.3554

$z_0 = \underline{1.06}$

d.

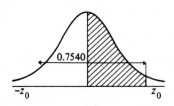

m (shaded area) = 0.3770

$z_0 = \underline{1.16}$

e.

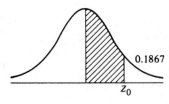

m (shaded area) = 0.3133

$z_0 = \underline{0.89}$

f.

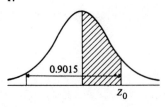

shaded area = 0.4015

$z_0 = \underline{1.29}$

SOLUTIONS: Chapter 6

13. To find the area under any normal curve (probability or proportion – see note after solution to Exercise 5), you must first convert the information to standard units. Once this conversion is complete, the problem then is similar to Exercises 1 through 11. Formula (6-3) is used to convert x units to standard units (z).

$$z = \frac{x - \mu}{\sigma} \qquad \mu = 100 \qquad \sigma = 10$$

a. $P(100 < x < 110) = P\left(\frac{100 - 100}{10} < z < \frac{110 - 100}{10}\right)$

$= P(0 < x < 1.0) = \underline{0.3413}$

b. $P(85 < x < 112) = P\left(\frac{85 - 100}{10} < z < \frac{112 - 100}{10}\right) = P(-1.5 < z < 1.2)$

$= 0.4332 + 0.3849 = \underline{0.8181}$

c. $P(x > 100) = P(z > 0.00) = \underline{0.5000}$

d. $P(x > 115) = P\left(z > \frac{115 - 100}{10}\right) = P(z > 1.5)$

$= 0.5000 - 0.4332 = \underline{0.0668}$

e. $P(105 < x < 130) = P\left(\frac{105 - 100}{10} < z < \frac{130 - 100}{10}\right)$

$= P(0.50 < z < 3.0) = 0.4987 - 0.1915 = \underline{0.3072}$

f. $P(75 < x < 135) = P\left(\frac{75 - 100}{10} < z < \frac{135 - 100}{10}\right)$

$= P(-2.50 < z < 3.50) = 0.4938 + 0.4998$

$= \underline{0.9936}$

g. $P(x > 110) = P\left(z > \frac{110 - 100}{10}\right) = P(z > 1.0) = 0.5000 - 0.3413$

$= \underline{0.1587}$

h. $P(x < 120) = P\left(z < \frac{120 - 100}{10}\right) = P(z < 2.00) = 0.5000 + 0.4772$

$= \underline{0.9772}$

14. See the comment in the solution to Exercise 13; $\mu = 9.3$, $\sigma = 2.1$.

a. $P(8.0 < x < 10.0) = P\left(\dfrac{8.0 - 9.3}{2.1} < z < \dfrac{10.0 - 9.3}{2.1}\right)$

$= P(-0.62 < z < 0.33) = 0.2324 + 0.1293$

$= \underline{0.3167}$

b. $P(10.0 < x < 14.5) = P\left(\dfrac{10.0 - 9.3}{2.1} < z < \dfrac{14.5 - 9.3}{2.1}\right)$

$= P(0.33 < z < 2.48) = 0.4934 - 0.1293$

$= \underline{0.3641}$

c. $P(5.0 < x < 8.0) = P\left(\dfrac{5.0 - 9.3}{2.1} < z < \dfrac{8.0 - 9.3}{2.1}\right)$

$= P(-2.05 < z < -0.62) = 0.4798 - 0.2324$

$= \underline{0.2474}$

d. $P(x > 7.5) = P\left(z > \dfrac{7.5 - 9.3}{2.1}\right) = P(z > -0.86) = 0.3051 + 0.5000$

$= \underline{0.8051}$

e. $P(x > 15.0) = P\left(z > \dfrac{15.0 - 9.3}{2.1}\right) = P(z > 2.71) = 0.5000 - 0.4966$

$= \underline{0.0034}$

f. $P(x < 12.0) = P\left(z < \dfrac{12.0 - 9.3}{2.1}\right) = P(z < 1.29) = 0.5000 + 0.4015$

$= \underline{0.9015}$

15. (1) The area under the normal curve between 10 and 25 is 0.85.
    (2) The proportion of the population whose value is between 10 and 25 is 0.85.
    (3) The probability that a randomly selected individual piece of data has a value between 10 and 25 is 0.85.

16. See the comment in the solution to Exercise 13.
    a. $\mu = 25.0$, $\sigma = 5.0$

$P(23.2 < x < 35.7) = P\left(\dfrac{23.2 - 25.00}{5.0} < z < \dfrac{35.7 - 25.0}{5.0}\right)$

$= P(-0.36 < z < 2.14) = 0.1406 + 0.4838$

$= \underline{0.6244}$

SOLUTIONS: Chapter 6

b. $\mu = 48.5, \sigma = 14.6$

$$P(56.2 < x < 92.1) = P\left(\frac{56.2 - 48.5}{14.6} < z < \frac{92.1 - 48.5}{14.6}\right)$$

$$= P(0.53 < z < 2.99) = 0.4986 - 0.2019$$

$$= \underline{0.2967}$$

c. $\mu = 128.9, \sigma = 6.2$

$$P(x > 134.5) = P\left(z > \frac{134.5 - 128.9}{6.2}\right) = P(z > 0.90)$$

$$= 0.5000 - 0.3159 = \underline{0.1841}$$

17. See the comments in the solutions to Exercises 13 and 15; $\mu = 68.5, \sigma = 3.2$.

a. $P(x < 62.0) = P\left(z < \frac{62.0 - 68.5}{3.2}\right) = P(z < -2.03)$

$$= 0.5000 - 0.4788 = \underline{0.0212}$$

b. $P(x > 72.0) = P\left(z > \frac{72.0 - 68.5}{3.2}\right) = P(z > 1.09)$

$$= 0.5000 - 0.3621 = \underline{0.1379}$$

c. The number of students who are shorter than 62 inches is found by multiplying the proportion by the number of students.

$$n(x < 62) = P(x < 62) \cdot 5000 = (0.0212)(5000) = \underline{106}$$

18. See the comments in the solutions to Exercises 13 and 15; $\mu = 155.6, \sigma = 7.9$.

a. $P(x < 150) = P\left(z < \frac{150 - 155.6}{7.9}\right) = P(z < -0.71)$

$$= 0.5000 - 0.2611 = \underline{0.2389}$$

b. $P(150 < x < 170) = P\left(\frac{150 - 155.6}{7.9} < z < \frac{170 - 155.6}{7.9}\right)$

$$= P(-0.71 < z < 1.82) = 0.2611 + 0.4656$$

$$= \underline{0.7267}$$

SOLUTIONS: Chapter 6

c. See comment in solution for Exercise 17c.

$$n(x > 170) = P(x > 170) \cdot 100 = P(z > 1.82) \cdot 100$$
$$= [0.5000 - 0.4656] \cdot 100 = (0.0344)(100) = 3.4 = \underline{3}$$

19. See the comments in the solutions to Exercises 13 and 15; $\mu = 114.4$, $\sigma = 17.2$.

a. $P(x > 128) = P\left(z > \dfrac{128 - 114.4}{17.2}\right) = P(z > 0.79)$

$= 0.5000 - 0.2852 = \underline{0.2148}$

b. $P(x < 80) = P\left(z < \dfrac{80 - 114.4}{17.2}\right) = P(z < -2.0)$

$= 0.5000 - 0.4772 = \underline{0.0228}$

c. $P(112 < x < 128) \cdot 500 = P\left(\dfrac{112 - 114.4}{17.2} < z < \dfrac{128 - 114.4}{17.2}\right) \cdot 500$

$= P(-0.14 < z < 0.79) \cdot 500$

$= [0.0557 + 0.2852] \cdot 500 = (0.3409)(500)$

$= 170.4 = \underline{170}$

20. a. 50%

In this problem the probability (percentage) is given and you are asked to work the formula relating z and x backwards to obtain a value for μ.

b.

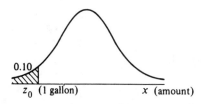

0.10 implies 0.4000 between z_0 and the mean; therefore, z_0 has the value −1.28.

$$-1.28 = \dfrac{1.000 - \mu}{0.015}$$

$\mu = 1.000 + (1.28)(0.015) = \underline{1.0192 \text{ gallons}}$

CHAPTER 7

SOLUTIONS TO EXERCISES 7-1

1. a. Population:

x	$P(x)$	$xP(x)$	$x^2P(x)$
0	1/4	0	0
2	1/4	2/4	4/4
4	1/4	4/4	16/4
6	1/4	6/4	36/4
		12/4 = 3	56/4 = 14

$\mu_x = \underline{3} \quad \sigma_x = \sqrt{14 - (3)^2} = \underline{\sqrt{5}}$

b. Distribution of samples:

Possible samples and their means:

```
00(0)      20(1)      40(2)      60(3)
02(1)      22(2)      42(3)      62(4)
04(2)      24(3)      44(4)      64(5)
06(3)      26(4)      46(5)      66(6)
```

c.

$\bar{x}$	$P(\bar{x})$	$\bar{x}P(\bar{x})$	$\bar{x}^2P(\bar{x})$
0	1/16	0	0
1	2/16	2/16	2/16
2	3/16	6/16	12/16
3	4/16	12/16	36/16
4	3/16	12/16	48/16
5	2/16	10/16	50/16
6	1/16	6/16	36/16
	16/16 = 1	48/16 = 3	184/16 = 11.5

d. $\mu_{\bar{x}} = \underline{3.0}; \; \sigma_{\bar{x}} = \sqrt{11.5 - (3.0)^2} = \sqrt{2.5} = \underline{\sqrt{5/2}}$

SOLUTIONS: Chapter 7

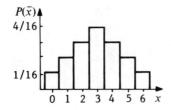

e. (1) $\mu_x = 3.0$ and $\mu_{\bar{x}} = 3.0$

(2) $\dfrac{\sigma_x}{\sqrt{n}} = \dfrac{\sqrt{5}}{\sqrt{2}}, \dfrac{\sigma_x}{\sqrt{n}} = \sqrt{\dfrac{5}{2}}$, and $\sigma_{\bar{x}} = \sqrt{\dfrac{5}{2}}$

(3) Histogram appears to be approximately normal.
All three properties are in agreement with the central limit theorem.

SOLUTIONS TO EXERCISES 7-2

1. The samples of size 2 were obtained by use of two-digit random numbers. The first digit represents the first element of the sample; the second digit, the second element. Each digit in the population is represented by 3 random digits, the tenth is ignored. (0 – 1,2,3; 2 – 4,5,6; 4 – 7,8,9; and 0 is ignored.) The following frequency distribution represents the 50 sample means obtained.

$\bar{x}$	f	$\bar{x}f$	$\bar{x}^2 f$
0	3	0	0
1	13	13	13
2	17	34	68
3	11	33	99
4	6	24	96
	50	104	276

$\bar{x} = \dfrac{104}{50} = \underline{2.08}$

$s_{\bar{x}} = \sqrt{\dfrac{(50)(276) - (104)(104)}{(50)(49)}} = \sqrt{1.218}$

$= \underline{1.10}$

Population:

x	$P(x)$	$xP(x)$	$x^2P(x)$
0	1/3	0	0
2	1/3	2/3	4/3
4	1/3	4/3	16/3
	1.0	6/3 = 2.0	20/3

$\mu_x = 2.0$

$\sigma_x = \sqrt{(20/3) - (2)^2} = \sqrt{8/3}$

$= 1.633$

Comparisons:

$\mu_x = \mu_{\bar{x}} = 2.0; \bar{\bar{x}} = 2.08$; approximately equal.

$\sigma_x = 1.633$, therefore $\sigma_x/\sqrt{n} = 1.633/\sqrt{2} = 1.155; s_{\bar{x}} = 1.10$.

These two values seem to support the central limit theorem. The distribution of frequencies (3, 13, 17, 11, 6) seems to be approximately normally distributed.

SOLUTIONS TO EXERCISES 7-3

1. $\mu = 26, \sigma = 4.5, n = 16$
   a. $P(24 < \bar{x} < 28) = P(-1.78 < z < +1.78) = 2(0.4625) = \underline{0.9250}$

   $z = \dfrac{28 - 26}{4.5/\sqrt{16}} = \dfrac{(2)(4)}{4.5} = 1.77\bar{7}$

   $z = \dfrac{24 - 26}{4.5/\sqrt{16}} = -1.77\bar{7}$

   b. $P(\bar{x} < 27) = P(z < 0.89) = 0.5000 + 0.3133 = \underline{0.8133}$

   $z = \dfrac{27 - 26}{4.5/\sqrt{16}} = \dfrac{4}{4.5} = 0.88\bar{8}$

   c. $P(25 < \bar{x} < 30) = P(-0.89 < z < 3.56) = 0.3133 + 0.4999 = \underline{0.8132}$

   $z = \dfrac{25 - 26}{4.5/\sqrt{16}} = -0.89$

   $z = \dfrac{30 - 26}{4.5/\sqrt{16}} = \dfrac{(4)(4)}{4.5} = 3.55\bar{5}$

2. $\mu = 62.4, \sigma = 10.8, n = 150$
   a. $P(62.4 - 0.75 < \bar{x} < 62.4 + 0.75) = P(-0.85 < z < 0.85)$

   $= 2(0.3023) = \underline{0.6046}$

SOLUTIONS: Chapter 7

$$z = \frac{\pm 0.75}{10.8/\sqrt{150}} = \pm 0.85$$

b. $P(\bar{x} < 63.0) = P(z < 0.68) = 0.5000 + 0.2517 = \underline{0.7517}$

$$z = \frac{63.0 - 62.4}{10.8/\sqrt{150}} = \frac{0.6\sqrt{150}}{10.8} = 0.68$$

3. $\mu = 30, \sigma = 6, n = ?$

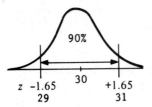

$P(29 < \bar{x} < 31) = 0.90$ and $P(-1.65 < z < 1.65) = 0.90$.

$$1.65 = \frac{31 - 30}{6/\sqrt{n}}$$

$$1.65 = \frac{\sqrt{n}}{6}$$

$\sqrt{n} = 6(1.65) = 9.90$

$n = (9.90)^2 = 98.01 = \underline{98}$

SOLUTIONS TO SELF-CORRECTING EXERCISES FOR CHAPTER 7

1. Note: These samples (2 through 20) were obtained by using the first five numbers in each column of Table 1.

   a. and b.

Sample		Σx	$\bar{x}$	Sample		Σx	$\bar{x}$
(1)	9,8,4,1,6	28	5.6	(11)	3,5,3,9,0	20	4.0
(2)	1,3,0,9,1	14	2.8	(12)	7,6,1,0,8	22	4.4
(3)	0,7,8,9,2	26	5.2	(13)	6,4,9,9,0	28	5.6
(4)	0,5,4,0,8	17	3.4	(14)	5,8,6,3,1	23	4.6
(5)	9,4,2,1,0	16	3.2	(15)	2,9,4,7,5	27	5.4
(6)	7,2,2,9,7	27	5.4	(16)	0,4,5,6,7	22	4.4
(7)	3,0,6,0,9	18	3.6	(17)	1,7,0,7,3	18	3.6
(8)	2,4,8,2,9	25	5.0	(18)	3,4,9,0,6	22	4.4
(9)	5,8,9,5,9	36	7.2	(19)	5,2,3,7,1	18	3.6
(10)	3,0,5,2,7	17	3.4	(20)	8,9,0,1,4	22	4.4

SOLUTIONS: Chapter 7

c. Means:

Class boundaries	Freq.
2.5 to 3.5	4
3.5 to 4.5	8
4.5 to 5.5	5
5.5 to 6.5	2
6.5 to 7.5	1
	20

d.

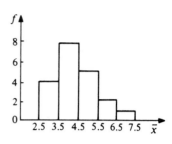

The set of 20 sample means listed above represents a proportion of the sampling distribution of sample means.

2. Recall that the median is the middle number when the sample is arranged in numerical order. For example, in sample 1 (1, 4, 6, 8, 9), 6 is the median.
   a. The medians for the 20 samples above are: 6, 1, 7, 4, 2, 7, 3, 4, 8, 3, 3, 6, 6, 5, 5, 5, 3, 4, 3, 4.
   b. Medians:

Class boundaries	Freq.
0.75 to 2.25	2
2.25 to 3.75	5
3.75 to 5.25	7
5.25 to 6.75	3
6.75 to 8.25	3
	20

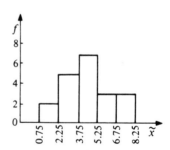

The set of 20 medians above represents a proportion of the sampling distribution of sample medians.

3. The midrange of each sample is obtained by adding the smallest and the largest valued pieces of data and dividing by 2. In sample 1, the midrange is $(1 + 9) \div 2 = 5.0$.
   a. The midrange values for the 20 samples above are: 5, 4.5, 4.5, 4, 4.5, 5.5, 4.5, 5.5, 7, 3.5, 4.5, 4, 4.5, 4.5, 5.5, 3.5, 3.5, 4.5, 4, 4.5.

SOLUTIONS: Chapter 7

b. Midranges:

Class boundaries	Freq.
2.25 to 3.75	3
3.75 to 5.25	13
5.25 to 6.75	3
6.75 to 8.25	1
	20

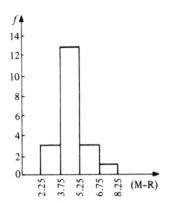

The set of 20 midranges above represents a proportion of the sampling distribution of sample midranges.

4. The range of each sample is obtained by subtracting the smallest valued piece of data from the largest valued piece of data (Range = $H - L$). In sample 1: Range = 9 - 1 = 8.
   a. The range values for these 20 samples are: 8, 9, 9, 8, 9, 7, 9, 7, 4, 7, 9, 8, 9, 7, 7, 7, 7, 9, 6, 9.
   b.

Range	Freq.
4	1
5	0
6	1
7	7
8	3
9	8
	20

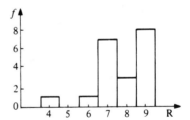

The set of 20 ranges above represents a proportion of the sampling distribution of sample ranges.

5. a. The standard deviation for each of these 20 samples can best be calculated by using the shortcut formula:

$$s = \sqrt{\frac{n(\Sigma x^2) - (\Sigma x)^2}{n(n-1)}}$$

The 20 values for the 20 samples above are: 3.21, 3.63, 3.96, 3.44, 3.56, 3.21, 3.91, 3.32, 2.05, 2.70, 3.32, 3.65, 3.78, 2.70, 2.70, 2.70, 3.29, 3.36, 2.41, 4.04.

SOLUTIONS: Chapter 7

b. Standard deviations:

Class boundaries	Freq.
2.00 to 2.50	2
2.50 to 3.00	4
3.00 to 3.50	7
3.50 to 4.00	6
4.00 to 4.50	1
	20

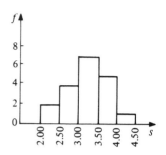

The set of 20 standard deviations above represents a proportion of the sampling distribution of sample standard deviations.

Note: In Exercises 1 through 5 we have seen sampling distributions of various sample statistics. Such a distribution is observed by obtaining any particular sample statistic for each of several repeated samples of a fixed size. The sampling distribution of sample means, $\bar{x}$, is the one that we are currently interested in and will study further in the following exercises.

6. a. The most convenient way to calculate the mean and the standard deviation of the 20 $\bar{x}$'s is to use the grouped frequency distribution obtained in Exercise 1c. Formulas (2-2) and (2-11) will be adapted for this purpose. In these formulas, n is 20, the number of values whose mean and standard deviation are being found.

Class $\bar{x}$ mark	f	$(\bar{x})f$	$(\bar{x})^2 f$
3.0	4	12	36
4.0	8	32	128
5.0	5	25	125
6.0	2	12	72
7.0	1	7	49
	20	88	410

$$\bar{\bar{x}} = \frac{\Sigma(\bar{x})f}{\Sigma f} = \frac{88}{20} = \underline{4.4}$$

$$s_{\bar{x}} = \sqrt{\frac{\Sigma(\bar{x})^2 f - (\Sigma \bar{x} f)^2}{(n-1)}}$$

$$= \sqrt{\frac{410 - \frac{(88)(88)}{20}}{(19)}}$$

$$= \sqrt{1.20} = 1.095$$

$$= \underline{1.1}$$

b. $\mu_{\bar{x}} = \mu = 4.5$; $\bar{\bar{x}} = 4.4$ from our observed set of 20 $\bar{x}$'s. 4.4 is close to 4.5; thus it agrees with the CLT.

c. $\sigma_{\bar{x}} = \sigma/\sqrt{n} = 2.87/\sqrt{5} = 1.28$; $s_{\bar{x}} = 1.1$ from our observed set of 20 $\bar{x}$'s. 1.1 is close to 1.28; thus it agrees with the CLT.

SOLUTIONS: Chapter 7

7. Each of these questions asks about one part of the CLT.
   a. $\mu_{\bar{x}} = \mu = 60$
   b. $\sigma_{\bar{x}} = \sigma/\sqrt{n} = 12/\sqrt{16} = \underline{3.0}$
   c. The distribution of $\bar{x}$'s is normal (bell-shaped).

8. Each of these questions asks about one part of the CLT.
   a. 78.0
   b. 1.25
   c. normal (bell-shaped)

9. The application of the CLT allows us to find probabilities associated with the sampling distribution of sample means by using the normal distribution. To use Table 5 for obtaining probabilities, the values of the $\bar{x}$ distribution must be converted to standard units using formula (7-2):

$$z = \frac{\bar{x} - \mu}{\sigma/\sqrt{n}}$$

After that, the probabilities are obtained by using Table 5 as shown in Chapter 6.

a. $P(\bar{x} > 48) = P\left(z > \dfrac{48 - 48}{8/\sqrt{25}}\right) = P(z > 0) = \underline{0.5000}$

b. $P(48 < \bar{x} < 51) = P\left(0 < z < \dfrac{51 - 48}{8/\sqrt{25}}\right) = P\left(0 < z < \dfrac{15}{8}\right)$

$= P(0 < z < 1.875) = \underline{0.4696}$

(halfway between 0.4693 and 0.4699)

c. $P(48 < \bar{x} < 53) = P\left(0 < z < \dfrac{53 - 48}{8/\sqrt{25}}\right) = P(0 < z < 3.125) = \underline{0.4991}$

d. $P(44 < \bar{x} < 48) = P\left(\dfrac{44 - 48}{8/\sqrt{25}} < z < 0\right) = P(-2.50 < z < 0) = \underline{0.4938}$

e. $P(44 < \bar{x} < 52) = P\left(\dfrac{44 - 48}{8/\sqrt{25}} < z < \dfrac{52 - 48}{8/\sqrt{25}}\right) = P(-2.50 < z < 2.50)$

$= 0.4938 + 0.4938 = \underline{0.9876}$

f. $P(\bar{x} < 52) = P\left(z < \dfrac{52 - 48}{8/\sqrt{25}}\right) = P(z < 2.50) = 0.5000 + 0.4938$

$= \underline{0.9938}$

SOLUTIONS: Chapter 7

10. See comment in Exercise 9 above; $\mu = 68.5$, $\sigma = 3.5$, $n = 25$.

a. $P(\bar{x} > 70) = P\left(z > \dfrac{70 - 68.5}{3.5/\sqrt{25}}\right) = P(z > 2.14) = 0.5000 - 0.4838$

$= \underline{0.0162}$

b. $P(67 < \bar{x} < 71) = P\left(\dfrac{67 - 68.5}{3.5/\sqrt{25}} < z < \dfrac{71 - 68.5}{3.5/\sqrt{25}}\right)$

$= P(-2.14 < z < 3.57) = 0.4838 + 0.4998$

$= \underline{0.9836}$

c. $P(\bar{x} > 68) = P\left(z > \dfrac{68.0 - 68.5}{3.5/\sqrt{25}}\right) = P(z > -0.71)$

$= 0.2611 + 0.5000 = \underline{0.7611}$

d. $P(\bar{x} < 67.5) = P\left(z < \dfrac{67.5 - 68.5}{3.5/\sqrt{25}}\right) = P(z < -1.43)$

$= 0.5000 - 0.4236 = \underline{0.0764}$

11. See comment in Exercise 9 above; $\mu = 64.50$, $\sigma = 11.80$, $n = 16$.

a. $P(\bar{x} > 60.00) = P\left(z > \dfrac{60.00 - 64.50}{11.80/\sqrt{16}}\right) = P(z > -1.53)$

$= 0.4370 + 0.5000 = \underline{0.9370}$

b. $P(65.00 < \bar{x} < 70.00) = P\left(\dfrac{65.00 - 64.50}{11.80/\sqrt{16}} < z < \dfrac{70.00 - 64.50}{11.80/\sqrt{16}}\right)$

$= P(0.17 < z < 1.86) = 0.4686 - 0.0675$

$= \underline{0.4011}$

c. $P(\bar{x} > 68.00) = P\left(z > \dfrac{68.00 - 64.50}{11.80/\sqrt{16}}\right) = P(z > 1.19)$

$= 0.5000 - 0.3830 = \underline{0.1170}$

d. $P(\bar{x} < 72.00) = P\left(z < \dfrac{72.00 - 64.50}{11.80/\sqrt{16}}\right) = P(z < 2.54)$

$= 0.5000 + 0.4945 = \underline{0.9945}$

e. $P(62.50 < \bar{x} < 66.50) = P\left(\dfrac{62.50 - 64.50}{11.80/\sqrt{16}} < z < \dfrac{66.50 - 64.50}{11.80/\sqrt{16}}\right)$

$= P(-0.68 < z < 0.68)$

$= 0.2517 + 0.2517 = \underline{0.5034}$

f. $P(\bar{x} < 65.50) = P\left(z < \dfrac{65.50 - 64.50}{11.80/\sqrt{16}}\right) = P(z < 0.34)$

$= 0.5000 + 0.1331 = \underline{0.6331}$

12. This exercise is an application of the CLT. See comment in Exercise 9 above; $\mu = 1250, \sigma = 75, n = 25$.

a. $P(\bar{x} > 1250) = P\left(z > \dfrac{1250 - 1250}{75/\sqrt{25}}\right) = P(z > 0.00) = \underline{0.5000}$

b. $P(\bar{x} > 1225) = P\left(z > \dfrac{1225 - 1250}{75/\sqrt{25}}\right) = P(z > -1.67)$

$= 0.4525 + 0.5000 = \underline{0.9525}$

c. $P(\bar{x} < 1300) = P\left(z < \dfrac{1300 - 1250}{75/\sqrt{25}}\right) = P(z < 3.33)$

$= 0.5000 + 0.4996 = \underline{0.9996}$

CHAPTER 8

SOLUTIONS TO EXERCISES 8-1

1. a. $H_0: p = P(\text{heads}) \leq 0.5$ (he cannot control getting heads)
   b. $H_a: p > 0.5$ (he can control getting heads)

   c.

   ```
              cannot control          |    can control
        +--+--+--+--+--+--+--+--+--+--+--+--+
        0  1  2  3  4  5  6  7  8    9  10 11 12
       (x)
                  decision point ——→
   ```

d. x is number of heads obtained on twelve flips.
e. If x is between 0 and 8, we will say that the friend lies; that is, he cannot control getting heads when he flips a penny. If x is 9 or more, we will agree that the friend seems to be able to control the results of flipping the penny.

2. a. $H_0: p = P(\text{makes basket}) \geq 0.80$ (he can make at least 80 percent of the free throws)
b. $H_a: p < 0.80$ (he makes less than 80 percent of the free throws)

c.

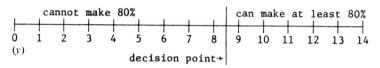

If we look at Table 4, we see that the probability for $x = 8$ is 0.121, while the probability for $x = 9$ is 0.054. Thus it seems reasonable to place the decision point between these two values of x.
d. y is number of free throws made during experiment of 14 attempts.
e. Looking at Table 4, if $p = 0.8$, then y-values of 8 or less are fairly rare events (less than a 0.05 chance). Thus if y is 8 or less, we will decide that the basketball player is not telling us the truth. If y is 9 or more, we will decide that he is apparently telling the truth.

3. a. $H_0: \mu = 3.50$ (mean hourly wage is $3.50)
b. $H_a: \mu \neq 3.50$ (mean hourly wage is not $3.50)
c. The test statistic to be used here is z, the mean hourly wage for the twenty students in the sample. The calculated value of z then would be compared to the scale shown in the figure below.

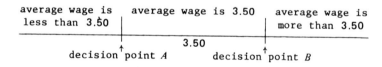

("Less than" and "more than" regions combine to make up the values that agree with H_a.) The values assigned to points A and B are somewhat arbitrary — they would be values that you believe to be distinctly different from 3.50. Perhaps 3.25 and 3.75 would be reasonable.
d. Select twenty students who work and ask their hourly wage. Calculate the sample mean.
e. Upon comparison of z to the scale shown above in the figure, we will decide to agree with either H_0 or H_a as described above.

SOLUTIONS TO EXERCISES 8-2

1. a. Type I error: a decision of "he can control the outcome" is reached, when in fact he has no talent at all.
   Type II error: a decision of "he cannot control the outcome" is reached when in fact he really does have some control over the outcome.
   b. $\alpha = P(x = 9, 10, 11,$ or $12,$ when $p = 0.5) = 0.054 + 0.016 + 0.003 + (0.0+)$
   $= \underline{0.073}$
   c. $\beta = P(x = 0, 1, 2, 3, 4, 5, 6, 7, 8,$ when $p = 0.7) = (0.0+) + (0.0+) + (0.0+) + 0.001 + 0.008 + 0.029 + 0.079 + 0.158 + 0.231 = \underline{0.506}$

2. a. $\alpha = P(x = 0, 1, 2, 3, 4, 5, 6, 7, 8,$ when $p = 0.80) = (0.0+) + (0.0+) + (0.0+) + (0.0+) + (0.0+) + (0.0+) + 0.002 + 0.009 + 0.032 = \underline{0.043}$
   b. $\beta = P(x = 9, 10, 11, 12, 13, 14,$ when $p = 0.5) = 0.122 + 0.061 + 0.022 + 0.006 + 0.001 + (0.0+) = \underline{0.212}$

3. a. $\alpha = P(x = 0, 1, 2, 3, 4, 5, 6, 7,$ when $p = 0.8) = (0.0+) + (0.0+) + (0.0+) + (0.0+) + 0.001 + 0.003 + 0.016 + 0.053 = \underline{0.073}$
   b. $\beta = P(x = 8, 9, 10, 11, 12,$ when $p = 0.50) = 0.121 + 0.054 + 0.016 + 0.003 + (0.0+) = \underline{0.194}$

SOLUTIONS TO EXERCISES 8-3

1. a. $z(0.05) = +1.65$
   b. $19.5 \pm 1.65 \cdot (12/\sqrt{16}) = 19.5 \pm (1.65)(3) = 19.5 \pm 4.95$
   (14.55 to 24.45)

2. a. Sample: $n = 25, \Sigma x = 105; \bar{x} = 4.2.$
   b. $4.2 \pm 1.15 \cdot (2.87/\sqrt{25}) = 4.2 \pm (1.15)(0.574) = 4.2 \pm 0.66$
   (3.54 to 4.86)
   c. If 100 different samples were to be drawn, one would expect that approximately 75 of the resulting confidence intervals would contain the true value of the population mean.

SOLUTIONS TO SELF-CORRECTING EXERCISES FOR CHAPTER 8

1. In determining the specific statements for the two hypotheses, you need to separate the three symbols (<, =, and >) into two groups. Remember, the group containing the equal sign becomes the null hypothesis.
   a. Mean age is 20.3 (=); mean age is not 20.3 ($\neq$).

   $H_0: \mu = 20.3$ $\qquad$ $H_a: \mu \neq 20.3$

SOLUTIONS: Chapter 8

b. Mean rental fee is at least $150 ($\geqslant$); mean rental fee is less than $150 ($<$).

$H_0: \mu = 150 (\geqslant)$ $\qquad$ $H_a: \mu < 150$

c. Mean family size is less than 4.3 ($<$); mean family size is greater than or equal to 4.3 ($\geqslant$).

$H_0: \mu = 4.3 (\geqslant)$ $\qquad$ $H_a: \mu < 4.3$

d. Mean distance is greater than 7.5 ($>$); mean distance is not greater than 7.5 ($\leqslant$).

$H_0: \mu = 7.5 (\leqslant)$ $\qquad$ $H_a: \mu > 7.5$

e. Mean time is 14.5 (=); mean time is not 14.5 ($\neq$).

$H_0: \mu = 14.5$ $\qquad$ $H_a: \mu \neq 14.5$

f. Mean amount is at least $42 ($\geqslant$); mean amount is less than $42 ($<$).

$H_0: \mu = 42 (\geqslant)$ $\qquad$ $H_a: \mu < 42$

2. $H_0: \mu = 150 (\geqslant)$ $\qquad$ (at least $150)
   $H_a: \mu < 150$ $\qquad$ (less than $150)

   a. Correct decision A: Fail to reject H_0 when H_0 is true. We decide to agree with statement "at least $150" when H_0 is true.
   b. Correct decision B: Reject H_0 when H_0 is false. We decide to reject the statement "at least $150" when H_0 is in fact false.
   c. Type I error: Reject H_0 when it is true. We decide not to agree with statement "at least $150" when it is in fact true.
   d. Type II error: Fail to reject H_0 when it is false. We decide to agree with claim "at least $150" when it is in fact false.
   e. Rejection of H_0 means we have evidence that the rent is less than $150.
   f. Failure to reject H_0 means we have evidence that the rent is at least $150.

3. $H_0: \mu = 7.5 (\leqslant)$ $\qquad$ (no more than 7.5)
   $H_a: \mu > 7.5$ $\qquad$ (greater than 7.5)

   a. We decide that the mean distance is no more than 7.5 miles and it is a correct decision.
   b. We decide that the mean distance is more than 7.5 miles and it is a correct decision.
   c. We decide that the mean distance is more than 7.5 miles when it is actually less than or equal to 7.5 miles.
   d. We decide that the mean is no more than 7.5 miles when it is actually greater than 7.5 miles.
   e. Fail to reject H_0 means that we conclude the mean distance is no more than 7.5 miles.
   f. Reject H_0 means that we conclude the mean distance is more than 7.5 miles.

SOLUTIONS: Chapter 8

4. Each of these z scores is found by converting the given probabilities (α) to the area between $z = 0$ and the desired z, which is then read from Table 5.
   a. area = 0.4000, $z = +1.28$
   b. area = 0.4750, $z = \pm 1.96$
   c. area = 0.4800, $z = +2.05$
   d. area = 0.4900 (left side), $z = -2.33$
   e. area = 0.4600, $z = +1.75$
   f. area = 0.4500, $z = \pm 1.65$
   g. area = 0.4500 (left side), $z = -1.65$
   h. area = 0.4900, $z = \pm 2.33$

5. The number in the parentheses is the area under the normal curve that lies to the right of the z being found. Each of these probabilities must be converted to the area between $z = 0$ and the z score being sought.
   a. area = 0.4500, $z(0.05) = +1.65$
   b. area = 0.4950, $z(0.005) = +2.575$ or $+2.58$
   c. area = 0.4750, $z(0.025) = +1.96$
   d. area = 0.4000, $z(0.10) = +1.28$
   e. area = 0.4600, $z(0.04) = +1.75$
   f. area = 0.4500 (left side), $z(0.95) = -1.645$ or -1.65
   g. area = 0.4900 (left side), $z(0.99) = -2.33$
   h. area = 0.4750 (left side), $z(0.975) = -1.96$
   i. area = 0.4800 (left side), $z(0.98) = -2.05$
   j. area = 0.4950 (left side), $z(0.995) = -2.575$ or -2.58

6. a. 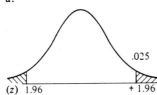 b.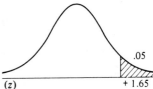

   c. 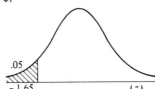 d.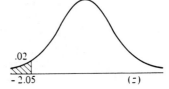

SOLUTIONS: Chapter 8

e.

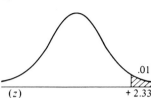

f.

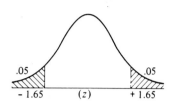

7. See Exercise 1c in reference to the null and alternative hypotheses. Formula (7-2) is used to calculate the value of the test statistic, z.

$H_0: \mu = 4.3$

$H_a: \mu \neq 4.3$

$\alpha = 0.05$

$z = \dfrac{\bar{x} - \mu}{\sigma/\sqrt{n}} = \dfrac{5.1 - 4.3}{1.0/\sqrt{20}}$

$z^* = +3.577 = 3.58$

Reject H_0: This data seems to indicate that μ is not equal to 4.3, since $z = 3.58$ falls in the rejection region.

8. The claim causes the information to be split into two parts: mean time is greater than 14.5 ($>$) and mean time is not greater than 14.5 ($\leq$). Formula (7-2) is used to calculate the value of the test statistic, z. $\bar{x} = 16.3$, $\sigma = 5.2$, $\alpha = 0.02$, $n = 100$.

$H_0: \mu = 14.5$ ($\leq$)

$H_a: \mu > 14.5$

$z = \dfrac{\bar{x} - \mu}{\sigma/\sqrt{n}} = \dfrac{16.3 - 14.5}{5.2/\sqrt{100}}$

$z^* = 3.46$

Reject H_0: The evidence supports the claim that college students' mean television watching time is greater than 14.5 hours per week.

9. This question calls for a confidence-interval estimate and therefore requires the use of formula (9-2).

$\bar{x} \pm z(\alpha/2)(\sigma/\sqrt{n})$

$138.50 \pm (1.65)(9.00/\sqrt{12})$

138.50 ± 4.29

$134.21 < \mu < 142.79$ (90% confidence interval)

10. See the comment in Exercise 9 above.

$$\bar{x} \pm z(\alpha/2)(\sigma/\sqrt{n})$$
$$115.0 \pm 1.96\,(8/\sqrt{50})$$
$$115.0 \pm 2.22$$

(112.8 to 117.2) the 95% confidence interval for μ

CHAPTER 9

SOLUTIONS TO EXERCISES 9-1

1. In inferences dealing with one value of the mean, the distinguishing characteristic between the use of z and t is whether or not σ is known. If σ is known, then we use the z statistic. If σ is unknown, then we always use the t statistic.

2. The standard error of the mean, $\sigma_{\bar{x}}$, is the standard deviation of the distribution of all possible $\bar{x}$-values that could be obtained for samples of a given size n. The standard error is equal in value to $\sigma/\sqrt{n}$, or is approximated by $s/\sqrt{n}$.

3. $t = \dfrac{\text{statistic} - \text{parameter}}{\text{estimate of standard error of statistic}}$

   $z = \dfrac{\text{statistic} - \text{parameter}}{\text{standard error of statistic}}$

4. statistic $\pm\, t(\text{df}, \alpha/2) \cdot$ (estimate of standard error of statistic), or
   statistic $\pm\, z(\alpha/2) \cdot$ (standard error of statistic)

SOLUTIONS TO EXERCISES 9-2

1. The standard error of proportion, $\sigma_{p'}$, is the standard deviation for the distribution of repeated values of p'. $\sigma_{p'}$ is found by use of the formula $\sqrt{pq/n}$ or is approximated by use of the formula $\sqrt{p'q'/n}$.

2. z statistic

SOLUTIONS: Chapter 9

SOLUTIONS TO EXERCISES 9-3

1. Chi square (χ^2) is used. Its pattern of distribution is shown in Figure 9-4 of the textbook.

SOLUTIONS TO SELF-CORRECTING EXERCISES FOR CHAPTER 9

1. To state the null hypothesis you will need to identify the population parameter under consideration and the value claimed. Second, you will need to split the set of three possible relationships ($<, =, >$) into two sets. The set containing the equal sign becomes the null hypothesis.

   a. "Standard deviation of lengths was no more than 4.5." $\sigma \leq 4.5$ or $\sigma > 4.5$.

   $$H_0: \sigma = 4.5 (\leq) \qquad H_a: \sigma > 4.5$$

   b. "Proportion who wear glasses is at least 40 percent." $p \geq 0.4$ or $p < 0.4$.

   $$H_0: p = 0.40 (\geq) \qquad H_a: p < 0.40$$

   c. "Percentage is seven." $p = 0.07$ or $p \neq 0.07$.

   $$H_0: p = 0.07 \qquad H_a: p \neq 0.07$$

   d. "Variance is less than 15." $\sigma^2 < 15$ or $\sigma^2 \geq 15$.

   $$H_0: \sigma^2 = 15 (\geq) \qquad H_a: \sigma^2 < 15$$

   e. "Percentage of foreign cars is at least 20 percent." $p \geq 0.20$ or $p < 0.20$.

   $$H_0: p = 0.20 (\geq) \qquad H_a: p < 0.20$$

   f. "Mean length is 19.2." $\mu = 19.2$ or $\mu \neq 19.2$.

   $$H_0: \mu = 19.2 \qquad H_a: \mu \neq 19.2$$

   g. "Variance was more than 150." $\sigma^2 > 150$ or $\sigma^2 \leq 150$.

   $$H_0: \sigma^2 = 150 (\leq) \qquad H_a: \sigma^2 > 150$$

2. These critical values are read directly from Table 6.
   a. ±2.14   b. +1.74   c. −2.47   d. ±2.76
   e. +2.78   f. −2.86   g. ±1.89   h. −1.65

3. The two numbers found in parentheses are the two values needed to locate each critical value of t. The first number is the number of degrees of freedom and the second number is the column heading (the area to the *right*). The critical values are read directly from Table 6.
   a. 1.72   b. 3.37   c. 2.98   d. 2.18
   e. 1.65   f. −1.75   g. −1.72   h. −2.33

4. Since σ is unknown in each of these situations, the test statistic will be t. The sign used in the alternative hypothesis (<, ≠, or >) indicates the location of the critical region. (<) points to the left tail, (>) points to the right tail, while (≠) means that both tails are critical regions. After that, each critical value is obtained from Table 6.

a.

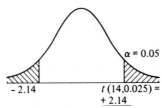

b.

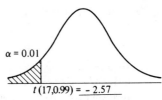

c.

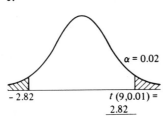

d.

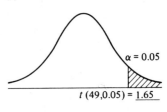

e.

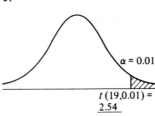

f.

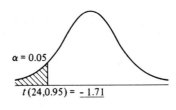

Note: H_0 about μ, σ unknown means you always use t.

5. These critical values of chi-square are obtained directly from Table 7. Remember that each column heading in the table is the area under the curve to the *right* of the value being sought.

a. 12.4, 39.4
b. 23.7
c. 7.26
d. 76.2
e. 1.24
f. 0.207, 14.9
g. 12.3, 33.9
h. 36.7

SOLUTIONS: Chapter 9

6. The two numbers found in parentheses are the two values needed to locate each critical value of chi-square. The first number is the number of degrees of freedom and the second number is the column heading (the area to the *right*).
   a. 37.7
   b. 13.6
   c. 9.24
   d. 30.2
   e. 13.2
   f. 6.00
   g. 7.26
   h. 8.23

7. See Exercises 4 and 5 in Chapter 8.
   a. +1.65
   b. ±1.96
   c. -2.05
   d. -1.28
   e. +1.65
   f. +2.05
   g. -2.33
   h. -1.96

8. Hypothesis tests dealing with p require the use of the z test statistic. The critical values are determined just as they were in Exercise 6, Chapter 8. Also, see the note in Exercise 4 above.

   a.
   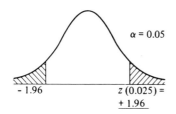

   b.
   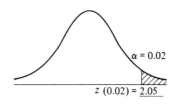

   c.
   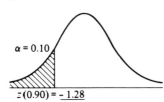

   When a hypothesis test deals with either variance (σ^2) or standard deviation (σ), the test statistic to be used is chi-square (χ^2). Otherwise the above guidelines apply.

   d.
   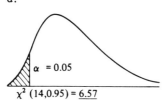

   e.
   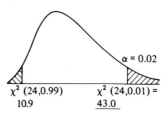

SOLUTIONS: Chapter 9

f.

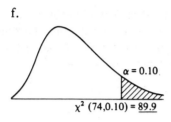

$\chi^2(74, 0.10) = 89.9$

This value of χ^2 must be found using interpolation.

$\chi^2(70, 0.10) = 85.5$
$\chi^2(74, 0.10) = 85.5 + 4.4 = \underline{89.9}$
$\chi^2(80, 0.10) = 96.6$

$\dfrac{4}{10} = \dfrac{x}{11.1}$

$x = 4.44$

9. a. Reject H_0.
   b. Fail to reject H_0.
   (See explanation of step 5 on pages 266 and 273 in the textbook.)

10. The parameter involved here is the mean (μ), since we are measuring the average American smoker by the number of cigarettes smoked per year. The test is one-tailed owing to the claim's use of the term "at least" ($\geq$). The test statistic to be used is Student's t, since the test deals with the mean, and the standard deviation of the population must be estimated by s. [t^* is calculated using formula (9-1).]

$H_0: \mu = 4{,}100 (\geq)$ (at least 4,100)
$H_a: \mu < 4{,}100$ (less than 4,100)

$t = \dfrac{\bar{x} - \mu}{s/\sqrt{n}}$

$t^* = \dfrac{4{,}015 - 4{,}100}{1{,}125/\sqrt{400}} = -1.51$

$\alpha = 0.05$, -1.65

Decision Fail to reject H_0.
Conclusion: There is no evidence to contradict the null hypothesis that the average American smoker appears to consume at least 4,100 cigarettes per year.

SOLUTIONS: Chapter 9

11. The mean cost is the value being sought. Therefore the 90 percent confidence interval will be calculated by using formula (9-2), since the estimate involves the mean, and the population standard deviation must be estimated using the standard deviation calculated from the sample.

$$\Sigma x = 492.50 \qquad \Sigma x^2 = 24402.75 \qquad n = 10$$

$$\bar{x} = 49.25 \qquad s = 4.04$$

$$1 - \alpha = 0.90$$

$$\bar{x} \pm t(9, 0.05)(s/\sqrt{n})$$

$$49.25 \pm (1.83)(4.04/\sqrt{10})$$

$$49.25 \pm 2.34$$

(46.91 to 51.59) 90% confidence interval for μ

12. This exercise deals with a hypothesis about a proportion (p). Therefore, the null hypothesis is $p = 0.95$, the percentage claimed by the statement being tested. The test statistic is therefore z and is obtained using formula (9-3).

$$H_0: p = 0.95 \qquad H_a: p \neq 0.95$$

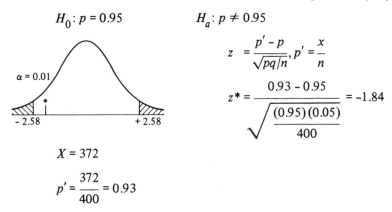

$$z = \frac{p' - p}{\sqrt{pq/n}}, \quad p' = \frac{x}{n}$$

$$z^* = \frac{0.93 - 0.95}{\sqrt{\frac{(0.95)(0.05)}{400}}} = -1.84$$

$$X = 372$$

$$p' = \frac{372}{400} = 0.93$$

Decision: Fail to reject H_0.
Conclusion: We cannot reject the claim that 95 percent of all honeymoons do not measure up to the brides' expectations.

13. This exercise asks for a confidence-interval estimation for the true proportion. Therefore, use formula (9-4).

$$n = 500 \qquad x = 61 \qquad 1 - \alpha = 0.95$$

$$p' = \frac{61}{500} = 0.122$$

$$p' \pm z(0.025)\sqrt{p'q'/n}$$

$$0.122 \pm 1.96 \sqrt{\frac{(0.122)(0.878)}{500}}$$

$$0.122 \pm 0.028$$

(0.094 to 0.150) 95% confidence interval for p

14. To test a claim about the standard deviation, we use the chi-square test statistic. The phrase "was greater than 12" causes the two sets of signs to be ($<, =$) and ($>$); therefore, the test is one-tailed and the critical region is on the right.

    $H_0: \sigma = 12\ (\leqslant)$ (was not greater than 12)
    $H_a: \sigma > 12$ (was greater than 12)
    $\alpha = 0.05$

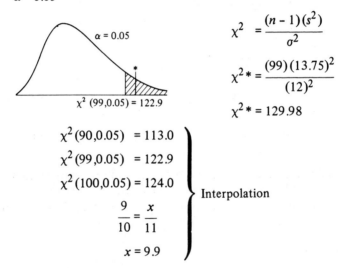

$$\chi^2 = \frac{(n-1)(s^2)}{\sigma^2}$$

$$\chi^2* = \frac{(99)(13.75)^2}{(12)^2}$$

$$\chi^2* = 129.98$$

$\chi^2(99,0.05) = 122.9$

$\left.\begin{array}{l}\chi^2(90,0.05) = 113.0 \\ \chi^2(99,0.05) = 122.9 \\ \chi^2(100,0.05) = 124.0 \\ \dfrac{9}{10} = \dfrac{x}{11} \\ x = 9.9\end{array}\right\}$ Interpolation

Decision: Reject H_0.
Conclusion: The standard deviation was significantly greater than 12.

15. The confidence-interval estimate for the variance of some variable is calculated using chi-square and formula (9-9), $1 - \alpha = 0.90$, $n = 20$, $\bar{x} = 119.5$, $s = 11.9$.

$$\frac{(n-1)s^2}{\chi^2(19,0.05)} < \sigma^2 < \frac{(n-1)s^2}{\chi^2(19,0.95)}$$

$$\frac{(19)(11.9)^2}{30.1} < \sigma^2 < \frac{(19)(11.9)^2}{10.1}$$

(89.4 to 266.4) 90% confidence interval for σ^2

CHAPTER 10

SOLUTIONS TO EXERCISES 10-1

1. The values of σ are both known and therefore the standard error is known — thus we use z, as we have before.

2. Yes, both formulas fit the mentioned forms:

    statistic: $\bar{x}_1 - \bar{x}_2$

    parameter: $\mu_1 - \mu_2$

    standard error of statistic: $\sqrt{\dfrac{\sigma_1^2}{n_1} + \dfrac{\sigma_2^2}{n_2}}$

SOLUTIONS TO EXERCISES 10-2

1. A *ratio* is used for comparing two variances or standard deviations, since relative size is more important than the difference.

2. The F statistic is used exclusively for inferences about the ratio of two variances.

3. Due to the use of three identifying values to determine a critical value, a different table is given for each value of α. The columns and rows of each table are then used for the degrees of freedoms for each of the two samples.

SOLUTION TO EXERCISES 10-3

1. Case A: both σ's known (Lesson 10-1)
    Cases B, C, D: Cases 1, 2, and 3 as described in Lesson 10-3 and shown on the tree diagram in Figure 10-2.

SOLUTIONS TO EXERCISES 10-4

1. The Student t statistic is used. The two dependent populations are paired and used to generate a third population. Actually the inferences are then made in reference to this third population, the paired differences. Even if both σ's were known, the standard deviation d would be unknown; therefore t is the only test statistic used.

2. Yes: statistic: $\bar{d}$, the mean of the differences

parameter: μ_d

estimate of standard error of $\bar{d}$: $s_d/\sqrt{n}$

SOLUTION TO EXERCISE 10-5

1. $\sqrt{p^*q^*\left(\dfrac{1}{n_1}+\dfrac{1}{n_2}\right)} = \sqrt{\dfrac{p^*q^*}{n_1}+\dfrac{p^*q^*}{n_2}}$

$= \sqrt{\left(\sqrt{\dfrac{p^*q^*}{n_1}}\right)^2 + \left(\sqrt{\dfrac{p^*q^*}{n_2}}\right)^2}$

$\uparrow \uparrow$
AB

Square roots A and B are the standard deviation for the p'_1 and p'_2. Square them and you have variance. Add them and then take the square root, and the result is the standard error for the new sampling distribution of $p'_1 - p'_2$

SOLUTIONS TO SELF-CORRECTING EXERCISES FOR CHAPTER 10

1. See comments for Self-Correcting Exercise 10-1.
   a. We are dealing with the difference between two proportions ($p_m - p_f$). The claim says "no difference," which is interpreted as a difference equal to zero.

   $H_0: p_m - p_f = 0 \qquad H_a: p_m - p_f \neq 0$

   b. The two variances are claimed to be equal ($\sigma_g^2 = \sigma_b^2$).

   $H_0: \sigma_g^2 = \sigma_b^2 \qquad H_a: \sigma_g^2 \neq \sigma_b^2$

   c. This claim deals with two population standard deviations and claims that $\sigma_m > \sigma_f$.

   $H_0: \sigma_m = \sigma_f (\leqslant) \qquad H_a: \sigma_m > \sigma_f$

   d. When dealing with two proportions, we typically work with their differences, namely ($p_b - p_r$). To say, "the proportion of blue is at least 10 percent greater" is to say that the difference ($p_b - p_r$) is 10 percent or more (as opposed to less than 10 percent).

   $H_0: p_b - p_r = 0.10 (\geqslant) \qquad H_a: p_b - p_r < 0.10$

SOLUTIONS: Chapter 10

e. This statement is about the difference between two independent means. The two graduating classes are separate groups and no difference between their mean value is claimed.

$$H_0: \mu_t - \mu_l = 0 \qquad\qquad H_a: \mu_t - \mu_l \neq 0$$

f. "Mean difference is positive." $\mu_d > 0$ or $\mu_d \leqslant 0$

$$H_0: \mu_d = 0(\leqslant) \qquad\qquad H_a: \mu_d > 0$$

g. "Average" must be equated with mean value. "Mean weight of American cars is 1,000 lb. more than mean of foreign cars," or their difference is 1,000 lb. $\mu_A - \mu_F = 1000$ or $\mu_A - \mu_F \neq 1000$.

$$H_0: \mu_A - \mu_F = 1000 \qquad\qquad H_a: \mu_A - \mu_F \neq 1000$$

2. (a.-f.) These values are obtained directly from Table 8a, 8b, or 8c in the textbook.
  a. $F(7, 11, 0.05) = 3.01$ \qquad b. $F(4, 7, 0.05) = 4.12$
  c. $F(15, 19, 0.025) = 2.62$ \qquad d. $F(24, 40, 0.025) = 2.01$
  e. $F(60, 24, 0.01) = 2.40$ \qquad f. $F(24, 60, 0.01) = 2.12$
  (g.-h.) Since the third number value in the parentheses is greater than 0.5, these two values are on the left-hand half of the F distribution. The use of formula (10-6) is required to determine the value being sought.

  g. $F(9, 10, 0.975) = \dfrac{1}{F(10, 9, 0.025)} = \dfrac{1}{3.96} = \underline{0.253}$

  $F(9, 10, 0.025) = \underline{3.78}$

  h. $F(15, 20, 0.99) = \dfrac{1}{F(20, 15, 0.01)} = \dfrac{1}{3.37} = \underline{0.297}$

  $F(15, 20, 0.01) = \underline{3.09}$

3. See the comments in Exercise 2 above. In addition to this, interpolation is necessary for parts d, e, and f.
  a. 3.69 \qquad b. 2.31 \qquad c. 2.02
  d. 2.045 (interpolated) \qquad e. 2.04 (interpolated) \qquad f. 1.69 (interpolated)

  g. $\dfrac{1}{4.06} = \underline{0.246}$ \qquad h. $\dfrac{1}{3.86} = \underline{0.259}$

4. Each of these is found by referring to the appropriate table in the back of the textbook.
  a. $z(0.05) = +1.65$ \qquad b. $\chi^2(9, 0.05) = 17.0$
  c. $F(9, 9, 0.05) = 3.18$ \qquad d. $t(9, 0.05) = 1.83$
  e. $\chi^2(14, 0.99) = 4.66$ \qquad f. $t(14, 0.99) = -[t(14, 0.01)] = -2.62$

g. $\pm[z(0.01)] = \pm 2.33$ h. $F(7, 7, 0.99) = \dfrac{1}{F(7, 7, 0.01)} = \dfrac{1}{6.99} = 0.143$

i. $\pm[t(24, 0.01)] = \pm 2.49$ j. $\chi^2(29, 0.10) = 39.1$

5. (a.-c.) Hypothesis tests dealing with the difference between two independent means (σ's known) use the z test statistic.

a.   b.

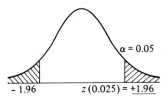
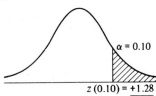

c.

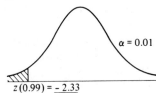

(d.-f.) When comparing the variances of two different populations, the F distribution is used. Remember: The location of the critical region is determined by the direction indicated in the alternative hypothesis.

d.   e.

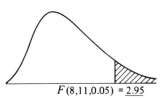
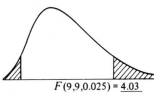

f.

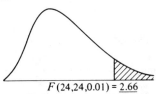

(g.-i.) When comparing the means from two independent samples where the standard deviations are unknown, we must use the Student's t distribution.

g.

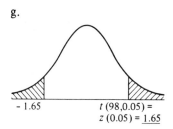

$t(98,0.05) = z(0.05) = \underline{1.65}$

h.

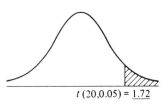

$t(20,0.05) = \underline{1.72}$

i.

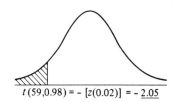

$t(59,0.98) = -[z(0.02)] = \underline{-2.05}$

(j.-l.) Hypothesis tests dealing with dependent means are handled in the same manner as hypothesis tests dealing with means with σ unknown (Exercise 4 in Self-Correcting Exercises, Chapter 9). The set of paired data gives rise to a set of paired differences (d_i). These values are then treated as one sample of values.

j.

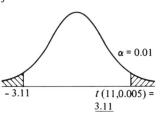

$t(11,0.005) = 3.11$

k.

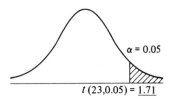

$t(23,0.05) = \underline{1.71}$

l.

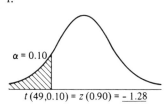

$t(49,0.10) = z(0.90) = \underline{-1.28}$

(m.-n.) The normal distribution is used to test all inferences concerning two proportions.

m.

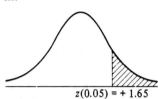

$z(0.05) = +1.65$

n.

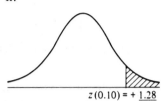

$z(0.10) = +1.28$

6. The difference between the two leagues is measured by the difference between the two mean values. Since the two leagues represent separate groups of people, we are dealing with independent samples and will use the z test statistic, since σ, the population standard deviation, is given. Use formula (10-1).

$H_0: \mu_s - \mu_{HD} = 0$
$H_a: \mu_s - \mu_{HD} \neq 0$

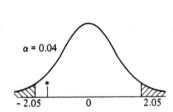

$\alpha = 0.04$

$-2.05 \quad 0 \quad 2.05$

$$z = \frac{(\bar{x}_s - \bar{x}_{HD}) - (0)}{\sqrt{\frac{(\sigma)^2}{n_s} + \frac{(\sigma)^2}{n_{HD}}}}$$

$$z^* = \frac{(158.3 - 160.7) - (0)}{\sqrt{\frac{(4.3)^2}{20} + \frac{(4.3)^2}{25}}}$$

$$= \frac{-2.40}{1.29}$$

$$= \underline{-1.86}$$

Decision: Fail to reject H_0.
Conclusion: There appears to be no significant difference between the mean bowling averages of these two leagues.

7. Samples of products produced by two separate manufacturers are independent samples. The difference between the mean values will be estimated using formula (10-2), since the two population standard deviations are given.

$$1 - \alpha = 0.95$$

$$(\bar{x}_A - \bar{x}_B) \pm z(\alpha/2) \sqrt{\frac{\sigma_A^2}{n_A} + \frac{\sigma_B^2}{n_B}}$$

$$(1450 - 1275) \pm 1.96 \sqrt{\frac{(120)^2}{150} + \frac{(90)^2}{100}}$$

SOLUTIONS: Chapter 10

$$175 \pm (1.96)(13.30)$$

$$175 \pm 26.1$$

(148.9 to 201.1) 95% confidence interval for $\mu_A - \mu_B$

8. The test is a comparison of standard deviations. The claim that the value now (σ_N) is greater than before determines the alternative hypothesis $(\sigma_N > \sigma_0)$.
$H_0: \sigma_N = \sigma_0 (\leq)$ N = recent 0 = two years ago
$H_a: \sigma_N > \sigma_0$
$\alpha = 0.05$

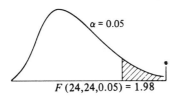

$$F = \frac{(s_N)^2}{(s_0)^2}$$

$$F* = \frac{(15.6)^2}{(10.6)^2} = \underline{2.16}$$

$F(24,24,0.05) = 1.98$

Decision: Reject H_0.
Conclusion: The standard deviation seems to have increased by a significant amount.

9. We have a claim about two independent means with both standard deviations (σ's) unknown.
The value of $z*$ will therefore be determined by use of formula (10-3).

$H_0: \mu_T - \mu_O = 0 (\leq)$
$H_a: \mu_T - \mu_O > 0$
$\alpha = 0.05$

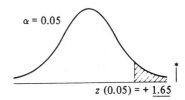

$z(0.05) = +1.65$

$$z = \frac{(\bar{x}_T - \bar{x}_O) - (\mu_T - \mu_O)}{\sqrt{\frac{(s_T)^2}{n_T} + \frac{(s_O)^2}{n_O}}} = \frac{(215.5 - 205.5) - (0)}{\sqrt{\frac{(15.6)^2}{76} + \frac{(17.6)^2}{190}}}$$

$$z* = \frac{10.0}{2.198} = \underline{+4.55}$$

Decision: Reject H_0.
Conclusion: The Top-Flite appears to average a longer distance than others when hit by a professional golfer.

SOLUTIONS: Chapter 10

10. To determine the correct formula to use, we need to test $\sigma_A = \sigma_B$ with an F test.

    $H_0: \sigma_A = \sigma_B$
    $H_a: \sigma_A \neq \sigma_B$

    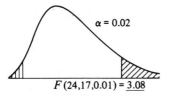

    $\alpha = 0.02$

    $F(24, 17, 0.01) = 3.08$

    $F^* = \dfrac{(13.6)^2}{(12.2)^2} = 1.24$

    Therefore, we can assume that $\sigma_A = \sigma_B$. Now use formula (10-13) to find the 98 percent confidence interval.

    $$(\bar{x}_A - \bar{x}_B) \pm t(41, 0.01) s_p \sqrt{\dfrac{1}{n_A} + \dfrac{1}{n_B}}$$

    $$(158.2 - 169.6) \pm (2.33) \sqrt{\dfrac{17(12.2)^2 + 24(13.6)^2}{18 + 25 - 2}} \sqrt{\dfrac{1}{18} + \dfrac{1}{25}}$$

    $-11.40 \pm (2.33)(13.04)(0.309)$

    -11.40 ± 9.39

    $\underline{(-20.79 \text{ to } -2.01)}$ 98% confidence interval for $\mu_A - \mu_B$

11. The effect of the drug is measured by the change in blood pressure (after and before). Thus we have paired data and need to test the difference between two dependent means. A one-tailed test will be used since the claim is that the drug will cause blood pressure to drop. Formula (10-17) will be used to find the test statistic t.

    $H_0: \mu_d = 0$ (does not drop)
    $H_a: \mu_d < 0$ (does drop)

    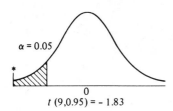

    $\alpha = 0.05$

    $t(9, 0.95) = -1.83$

    $t = \dfrac{\bar{d} - \mu_d}{s_d/\sqrt{n}}$

    $t^* = \dfrac{-1.7 - 0}{1.49/\sqrt{10}}$

    $= \underline{-3.6}$

    Decision: Reject H_0.
    Conclusion: The drug seems to cause a significant drop in blood pressure.

SOLUTIONS: Chapter 10

12. This is a difference-of-two-dependent-means problem, with the paired data being prices at specific stations on two different dates. The confidence-interval estimate is made by using formula (10-18).

    d = Dec. price–Aug. price
    d = 10.5, 9.0, 6.0, 6.2, 12.0, 8.6, 7.2, 7.2, 12.8, 4.0, 5.0, 2.0, 4.0, 6.0, 4.0
    $\Sigma d = 104.5, \Sigma d^2 = 864.17, \bar{d} = 6.97, s_d = 3.12$
    $1 - \alpha = 0.95$

    $$\bar{d} \pm t(14, 0.025)(s_d/\sqrt{n})$$

    $$6.97 \pm (2.14)(3.12/\sqrt{15})$$

    $$6.97 \pm 1.72$$

    (5.25 to 8.69) 95% confidence interval for μ_d

13. This hypothesis test deals with the difference between two proportions; the proportion of men who responded no and the proportion of women who responded no.

    $H_0: p_m - p_f = 0$
    $H_a: p_m - p_f \neq 0$

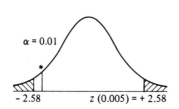

$\alpha = 0.01$

$-2.58 \quad z(0.005) = +2.58$

$x_m = 358$
$x_f = 372$
$n_m = n_f = 400$
$p_m' = \dfrac{358}{400} = 0.895$
$p_f' = \dfrac{372}{400} = 0.930$

$$z = \frac{(p_m' - p_f') - (p_m - p_f)}{\sqrt{p^*q^*\left(\dfrac{1}{n_m} + \dfrac{1}{n_f}\right)}}$$

$p^* = \dfrac{358 + 372}{400 + 400} = \dfrac{730}{800}$

$p^* = 0.9125 \quad q^* = 0.0875$

$$z^* = \frac{(0.895 - 0.930) - (0)}{\sqrt{(0.9125)(0.0875)\left(\dfrac{1}{400} + \dfrac{1}{400}\right)}}$$

$$= \frac{-0.035}{0.01997} = -1.75$$

Decision: Fail to reject H_0.
Conclusion: There is no significant difference found between these sets of sample results.

SOLUTIONS: Chapter 10

14. Like Exercise 13 above, this hypothesis test deals with the difference between two proportions.

$H_0: p_b - p_g = 0 \; (\leq)$
$H_a: p_b - p_g > 0$

$x_b = 288, n_b = 600$
$x_g = 125, n_g = 500$

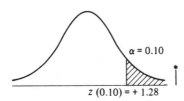

$\alpha = 0.10$

$z(0.10) = +1.28$

$p_b' = \dfrac{288}{600} = 0.48$

$p_g' = \dfrac{125}{500} = 0.25$

$z = \dfrac{(p_b' - p_g') - (p_b - p_g)}{\sqrt{p^*q^*\left(\dfrac{1}{n_b} + \dfrac{1}{n_g}\right)}}$

$p^* = \dfrac{288 + 125}{600 + 500} = \dfrac{413}{1100}$

$p^* = 0.375, q^* = 0.625$

$z^* = \dfrac{(0.48 - 0.25) - (0)}{\sqrt{(0.375)(0.625)\left(\dfrac{1}{600} + \dfrac{1}{500}\right)}} = \dfrac{0.230}{0.029} = 7.93$

Decision: Reject H_0.
Conclusion: It appears that boys are more likely to commit the destruction-of-property offense than are girls.

15. In order to estimate the difference between two proportions we will use the z statistic and formula (10-21).
a. $1 - \alpha = 0.95$

$p^* = \dfrac{x_3 + x_4}{n_3 + n_4}$

$x_3 = (0.057)(1000) = 57$
$x_4 = (0.060)(1200) = 72$

$p^* = \dfrac{57 + 72}{1000 + 1200} = \dfrac{129}{2200} = 0.06$

$(p_4' - p_3') \pm z(0.025) \sqrt{(p^*)(q^*)\left(\dfrac{1}{n_3} + \dfrac{1}{n_4}\right)}$

$(0.060 - 0.057) \pm (1.96) \sqrt{(0.06)(0.94)\left(\dfrac{1}{1000} + \dfrac{1}{1200}\right)}$

$0.003 \pm (1.96)(0.0101)$

0.0030 ± 0.0199

$(\underline{-0.0169 \text{ to } 0.0229})$ 95% confidence interval for $p_4 - p_3$

b. No, it does not. The confidence interval contains zero, thus the evidence does not suggest a significant change in the percentage of people who are saving.

CHAPTER 11

SOLUTIONS TO EXERCISES 11-1

1.

	Observed	Expected	$\frac{(O-E)^2}{E}$
Odd	107	100	0.49
Even	93	100	0.49
Total			0.98

H_0: Odd and even digits occurred with equal frequency.

$df = k - 1 = 2 - 1 = 1$

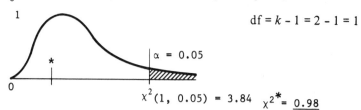

$\chi^2(1, 0.05) = 3.84 \qquad \chi^{2*} = 0.98$

Decision: Fail to reject H_0.
Conclusion: There is no evidence to contradict that the odd and even digits appear to be equally frequent.

2. These tests will follow the format of the illustration and Exercise 1 of this lesson. (Each sample drawn will be different — thus no solution is shown here.)

SOLUTIONS TO EXERCISES 11-2

1.

32 (39.3)	35 (27.7)	67
42 (34.7)	17 (24.3)	59
74	52	126

SOLUTIONS: Chapter 11

a. H_0: Response is independent of sex.

b.

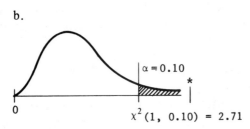

$x^2(1, 0.10) = 2.71$

c.

	Head	Tail	
Male	32 [39.35]	35 [27.65]	67
Female	42 [34.65]	17 [24.35]	59
	74	52	126

$$x^{2*} = \frac{(32 - 39.35)^2}{39.35} + \frac{(35 - 27.65)^2}{27.65} + \frac{(42 - 34.65)^2}{34.65} + \frac{(17 - 24.35)^2}{24.35}$$

$= 7.1$

d. Decision: Reject H_0.
e. Conclusion: The response does not seem to be independent of the sex of the respondent.

SOLUTIONS TO SELF-CORRECTING EXERCISES FOR CHAPTER 11

1. a. Yes, it fits the form of the multinomial experiment. Frequencies are the number of people in each of several categories.
   b. Yes, it fits the form of a multinomial experiment. The number of garments (frequency) in each category is reported.
   c. Assuming the classification according to grade-point average is in the form of intervals, we would have data, frequencies, fitting the form of a contingency table.
   d. Yes, multinomial. (Similar to parts a and b.)
2. a. Each outcome of a trial results in a response that fits into exactly one of k categories. Each trial is independent of the others and only the frequency (number of elements) for the category is recorded.

b. Each cell must have an expected frequency of five or more. As a check, one should always be sure that the total of the expected frequencies is exactly the same as the total of the observed frequencies.
c. A contingency table results from information being categorized according to two different sets of characteristics, thus causing the chart displaying the resulting frequencies to be composed of rows and columns.
d. Each cell must have an expected frequency of five or more. The total of the expected values, column totals, row totals, and grand total, must be exactly equal to the corresponding observed totals.

3. The claim that the number of arrests per week is constant is consistent with the idea of equal expected frequencies. Each expected frequency value is therefore one-tenth of the total number of arrests (100). Thus $E = 10$ for each cell in the following chart.
H_0: The number of arrests per week is constant.

| O | 12 | 9 | 20 | 3 | 12 | 10 | 15 | 7 | 8 | 4 | 100 |
| E | 10 | 10 | 10 | 10 | 10 | 10 | 10 | 10 | 10 | 10 | 100 |

| O - E | 2 | -1 | 10 | -7 | 2 | 0 | 5 | -3 | -2 | -6 | 0 |
| $\frac{(O - E)^2}{10}$ | .4 | .1 | 10.0 | 4.9 | .4 | 0 | 2.5 | .9 | .4 | 3.6 | 23.2 |

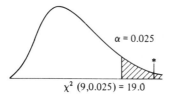

$$\chi^{2*} = \underline{23.2}$$

Decision: Reject H_0.
Conclusion: These observed frequencies do not appear to be uniformly distributed on a weekly basis.

4. a. H_0: P(milk chocolate) = P(soft nougat) = P(crunchy caramel) = 1/3
   b. Each expected frequency is equal to 1/3(75).

				totals
O	15	29	31	75
E	25	25	25	75

| O - E | -10 | 4 | 6 | 0 (ck) |
| $\frac{(O - E)^2}{25}$ | 4.00 | 0.64 | 1.44 | 6.08 |

$$\chi^{2*} = \underline{6.08}$$

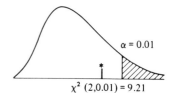

Decision: Fail to reject H_0.
Conclusion: At the 0.01 level of significance, there is no evidence to reject the claim that the three kinds of candy bar are equally popular among these nursery school children.

5. H_0: Nondominant genes $P(\text{red}) = 1/4, P(\text{pink}) = 1/2, P(\text{white}) = 1/4$.

	red	pink	white	totals
Observed	112	271	117	500
Expected	125	250	125	500

$E(\text{red}) = (1/4)(500), E(\text{pink}) - (1/2)(500), E(\text{white}) = (1/4)(500)$

$$\chi^{2*} = \frac{(112-125)^2}{125} + \frac{(271-250)^2}{250} + \frac{(117-125)^2}{125}$$

$= 1.35 + 1.76 + 0.51$

$= \underline{3.62}$

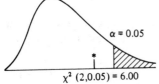

$\chi^2(2, 0.05) = 6.00$

Decision: Fail to reject H_0.
Conclusion: The experimental set of data does not contradict the idea of nondominant genes in the Four O'Clock flowers.

6. a. This experiment involves the binomial distribution in that each cell could be thought of as representing x, the number of males out of a set of four children. [$n = 4$ (number of trials), $p = P$(male child on any one birth).]
   b. H_0: Binomial distribution with $n = 4$ and $p = 1/2$.

x	P(x)	O	E	(O − E)²/E
4	0.062	25	31.0	1.161
3	0.250	107	125.0	2.592
2	0.375	197	187.5	0.481
1	0.250	143	125.0	2.592
0	0.062	28	31.0	0.290
	0.999	500	499.5	7.116

(round-off error)

$\chi^2 = \underline{7.116}$

Decision: Fail to reject H_0.

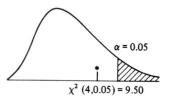

Conclusion: The observed frequencies are not significantly different from the frequencies that would be expected in a binomial distribution where $n = 4$ and $p = 1/2$. Therefore, the evidence supports the null hypothesis that male and female births are equally likely.

7. H_0: The sex of subject and the preference made are independent.

	A	B	C	D	E	totals
Male	60 \newline 71.79	33 \newline 41.68	51 \newline 38.21	32 \newline 18.53	0 \newline 5.79	176
Female	126 \newline 114.21	75 \newline 66.32	48 \newline 60.79	16 \newline 29.47	15 \newline 9.21	280
Totals	186	108	99	48	15	456

Note: Each of the expected frequencies above were calculated using formula (13-3). $71.79 = 186 \times 176/456$, $41.68 = 108 \times 176/456$, etc.

$$\chi^{2*} = \frac{(60 - 71.79)^2}{71.79} + \frac{(33 - 41.68)^2}{41.68} + \frac{(51 - 38.21)^2}{38.21} + \frac{(32 - 18.53)^2}{18.53}$$
$$+ \frac{(0 - 5.79)^2}{5.79} + \frac{(126 - 114.21)^2}{114.21} + \frac{(75 - 66.32)^2}{66.32} + \frac{(48 - 60.79)^2}{60.79}$$
$$+ \frac{(16 - 29.47)^2}{29.47} + \frac{(15 - 9.21)^2}{9.21}$$

$\chi^{2*} = 1.936 + 1.809 + 4.281 + 9.799 + 5.789 + 1.217 + 1.137 + 2.691$
$\quad + 6.159 + 3.639$

$\chi^{2*} = \underline{38.457}$

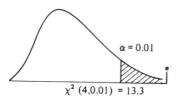

Decision: Reject H_0.

Conclusion: The evidence suggests that the subject's sex has a significant effect on the preference made.

8. a. H_0: P(passing) is the same in all of the instructors' classes
or
The students' passing or failing is independent of the instructor.

b.

	W	X	Y	Z	totals
Passed	51 | 46.68	46 | 49.10	63 | 58.76	38 | 43.46	198
Failed	7 | 11.32	15 | 11.90	10 | 14.24	16 | 10.54	48
Totals	58	61	73	54	246

(See note under table of Exercise 7.)

$$\chi^{2*} = \frac{(51 - 46.68)^2}{46.68} + \frac{(46 - 49.10)^2}{49.10} + \ldots$$

$= 0.399 + 0.195 + 0.307 + 0.687 + 1.647 + 0.806 + 1.264$

$+ 2.833$

$\chi^{2*} = \underline{8.138}$

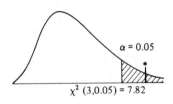

$\alpha = 0.05$

$\chi^2 (3, 0.05) = 7.82$

Decision: Reject H_0.
Conclusion: It appears that a passing or failing grade is not independent of the instructor.

SOLUTIONS: Chapter 11

9. H_0: The proportion of above-average, average, and below-average grades is the same for each year.
or
The distribution of grades is independent of the year.

	year				
	1	2	3	4	
(A or B)	38 ⌊31.3⌋	28 ⌊32.1⌋	32 ⌊26.1⌋	18 ⌊26.5⌋	116
(C)	18 ⌊23.2⌋	23 ⌊23.8⌋	21 ⌊19.4⌋	24 ⌊19.6⌋	86
(D or F)	22 ⌊23.5⌋	29 ⌊24.1⌋	12 ⌊19.5⌋	24 ⌊19.9⌋	87
	78	80	65	66	289

(See note for table of Exercise 7.)

$$\chi^{2*} = \frac{(38 - 31.3)^2}{31.3} + \frac{(28 - 32.1)^2}{32.1} + \ldots$$

$$\chi^{2*} = 1.43 + 0.53 + 1.34 + 2.72 + 1.17 + 0.03 + 0.14 + 0.97 + 0.09$$
$$+ 1.00 + 2.93 + 0.86$$

$$\chi^{2*} = \underline{13.21}$$

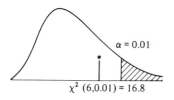

$\chi^2 (6, 0.01) = 16.8$

Decision: Fail to reject H_0.
Conclusion: At the 0.01 level of significance, the evidence does not show sufficient strength to reject the independence claim.

SOLUTIONS: Chapter 11

10. a. H_0: Political party identity is distributed the same for each of the calendar years. (This is a test of homogeneity.)

	Rep.	Dem.	Ind.	Totals
1950	33 | 29	45 | 44.75	22 | 26.75	100
1960	30 | 29	47 | 44.75	23 | 26.75	100
1970	29 | 29	45 | 44.75	26 | 26.75	100
1974	24 | 29	42 | 44.75	34 | 26.75	100
Totals	116	179	105	400

$$\chi^{2*} = \frac{(33-29)^2}{29} + \frac{(45-44.75)^2}{44.75} + \frac{(22-26.25)^2}{26.25} + \ldots$$

$\chi^{2*} = 0.552 + 0.001 + 0.688 + 0.034 + 0.113 + \ldots$

$\chi^{2*} = 5.114$

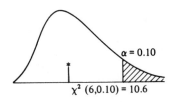

$\chi^2 (6, 0.10) = 10.6$

Decision: Fail to reject H_0.

Conclusion: Using a sample size of 100 per year and a 0.10 level of significance, we can not reject the "same distribution" null hypothesis.

b.

	Rep.	Dem.	Ind.	Totals
1950	66 |58	90 |89.5	44 |52.5	200
1960	60 |58	94 |89.5	46 |52.5	200
1970	58 |58	90 |89.5	52 |52.5	200
1974	48 |58	84 |89.5	68 |52.5	200
Totals	232	358	210	800

$$\chi^{2*} = \frac{(66-58)^2}{58} + \frac{(90-89.5)^2}{89.5} + \frac{(44-52.5)^2}{52.5} + \ldots$$

$$\chi^{2*} = 1.103 + 0.002 + 1.376 + 0.069 + 0.226 + \ldots$$

$$\chi^{2*} = 10.22$$

Comment: Everything doubled.

c. If the sample size were to double again, it seems that everything else would double also. Eventually the χ^{2*} will become larger than the critical value and cause H_0 to be rejected, while the percentages remain intact.

CHAPTER 12

SOLUTIONS TO EXERCISES 12-1

1. $s^2 = \dfrac{\Sigma(x-\bar{x})^2}{n-1} = \dfrac{n(\Sigma x^2) - (\Sigma x)^2}{n(n-1)} = \dfrac{\dfrac{n(\Sigma x^2) - (\Sigma x)^2}{n}}{n-1}$

SOLUTIONS: Chapter 12

Therefore:

$$\Sigma(x-\bar{x})^2 = \frac{n(\Sigma x^2)-(\Sigma x)^2}{n}$$

$$\Sigma(x-\bar{x})^2 = \Sigma x^2 - \frac{(\Sigma x)^2}{n}$$

2.

A		B		C		
x	x^2	x	x^2	x	x^2	
3	9	5	25	8	64	$\Sigma x = 60$
2	4	6	36	7	49	$\Sigma x^2 = 338$
4	16	5	25	6	36	
3	9	4	16	7	49	
12	38	20	102	28	198	

a. $SS(total) = 338 - \frac{(60)^2}{12} = 338 - \frac{3600}{12} = \underline{38}$

$SS(error) = 338 - \left[\frac{(12)^2}{4} + \frac{(20)^2}{4} + \frac{(28)^2}{4}\right] = 338 - 322 = \underline{6}$

$SS(factor) = 332 - 300 = \underline{32}$

b. $df(total) = 12 - 1 = \underline{11}$

$df(factor) = 3 - 1 = \underline{2}$

$df(error) = 3(4-1) = \underline{9}$

$MS(factor) = \frac{32}{2} = \underline{16.0}$

$MS(error) = \frac{6}{9} = \underline{0.67}$

ANOVA Table:

Source	SS	df	MS	
Factor	32	2	16.0	$F^* = 24.0$
Error	6	9	0.67	
Total	38	11		

c. $\bar{x} = \frac{\Sigma x}{n} = \frac{60}{12} = 5$

SOLUTIONS: Chapter 12

$\Sigma(x - \bar{x})^2 = [(-2)^2 + (-3)^2 + (-1)^2 + (-2)^2 + 0 + (1)^2 + 0 + (-1)^2 + (+3)^2 +$
$\qquad\qquad\qquad (+2)^2 + (+1)^2 + (+2)^2]$
$= 4 + 9 + 1 + 4 + 1 + 1 + 9 + 4 + 1 + 4$
$= 38$

They are the same, 38 each.

d. $(s_p)^2 = \left[\dfrac{(n_A - 1)s_A^2 + (n_B - 1)s_B^2 + (n_C - 1)s_C^2}{n_A + n_B + n_C - 3}\right]$

$(s_p)^2 = \left[\dfrac{3(2/3) + 3(2/3) + 3(2/3)}{4 + 4 + 4 - 3}\right] = \dfrac{6}{9} = 0.67$

(1) The numerator for $(s_p)^2$ is the same as SS (error).
(2) The denominator for $(s_p)^2$ is the same as df (error).
(3) $(s_p)^2$ and MS (error) are equal.

e. $H_0: \sigma_{\mu_r}^2 = 0$ (factor has no effect)
   $H_a: \sigma_{\mu_r}^2 > 0$ (factor has an effect)

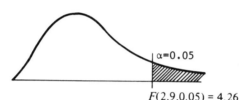

$F(2, 9, 0.05) = 4.26$

$F* = \dfrac{16.0}{0.67} = 24.0$

Decision: Reject H_0.
Conclusion: The tested factor does seem to have an effect on the process in question.

SOLUTIONS TO SELF-CORRECTING EXERCISES FOR CHAPTER 12

1. a. The subscript of an individual x, cr, identifies first the column in which the piece of data lies and second its position within that column. Therefore, the subscript 12 means column 1, the second piece of data.

$x_{12} = 3 \qquad x_{41} = 6 \qquad x_{52} = 7$
$x_{31} = 4 \qquad x_{23} = 8 \qquad x_{42} = 5$

SOLUTIONS: Chapter 12

b. The subscript used with a column total, C_k, identifies the column number. Thus C_1 is the sum of all the values in the first column.

$C_1 = 5$ $C_2 = 9$ $C_3 = 6$
$C_4 = 13$ $C_5 = 8$ $T = 41$

c. $\Sigma x = 41$ (summation of all x's — grand total T)

$\Sigma(x^2) = 211$ (sum of all x^2)

$\Sigma C_k = 41$ (sum of column totals — same as grand total)

d. $k = 3$ (number of replicates per column).

$n = 15$ (number of pieces of data altogether)

$c = 5$ (number of columns)

2. a. $x_{11} = 7$ $x_{13} = 3$ $x_{21} = 4$ $x_{24} = 5$
$x_{35} = 6$ $x_{46} = 8$ $x_{31} = 9$ $x_{42} = 2$

b. $C_1 = 12$ $C_3 = 18$ $C_4 = 13$ $T = 54$

c. $\Sigma x = T = 54$ $\Sigma(x^2) = 294$

d. $k = 6$ $n = 24$ $c = 4$

(See Exercise 1.)

3. a. $SS(\text{total}) = \Sigma(x^2) - \dfrac{(T)^2}{n}$

$= 211 - \dfrac{(41)^2}{15} = 211 - 122.066$

$= \underline{98.93\overline{3}}$

$SS(\text{factor}) = \left[\dfrac{C_1^2}{k_1} + \dfrac{C_2^2}{k_2} + \ldots\right] - \dfrac{(T)^2}{n}$

$= \left[\dfrac{5^2}{3} + \dfrac{9^2}{3} + \dfrac{6^2}{3} + \dfrac{13^2}{3} + \dfrac{8^2}{3}\right] - 122.06\overline{6} = 125 - 122.06\overline{6}$

$= \underline{12.93\overline{3}}$

$SS(\text{error}) = \Sigma(x^2) - \left[\dfrac{C_1^2}{k_1} + \dfrac{C_2^2}{k_2} + \ldots\right]$

$= 211 - 125$

$= \underline{86.0}$

SOLUTIONS: Chapter 12

b. df(total) = $n - 1$ = 15 - 1 = $\underline{14}$
df(factor) = $c - 1$ = 5 - 1 = $\underline{4}$
df(error) = $n - c$ = 15 - 5 = $\underline{10}$
[df(factor) + df(error) = df(total) (ck)]

c. MS(factor) = $\dfrac{SS(factor)}{df(factor)} = \dfrac{12.93\overline{3}}{4} = \underline{3.2\overline{3}}$

MS(error) = $\dfrac{SS(error)}{df(error)} = \dfrac{86.0}{10} = \underline{8.6}$

d. $F^* = \dfrac{MS(factor)}{MS(error)} = \dfrac{3.2\overline{3}}{8.6} = \underline{0.376}$

4. a. SS(total) = $294 - \dfrac{(54)^2}{24}$ (See formula in Exercise 3.)

= 294 - 121.5 = $\underline{172.5}$

SS(factor) = $\left[\dfrac{12^2}{6} + \dfrac{11^2}{6} + \dfrac{18^2}{6} + \dfrac{13^2}{6}\right] - 121.5$

= 126.3$\overline{3}$ - 121.5 = $\underline{4.83\overline{3}}$

SS(error) = 294 - 126.3$\overline{3}$ = $\underline{167.6\overline{6}}$

b. df(total) = 24 - 1 = $\underline{23}$
df(factor) = 4 - 1 = $\underline{3}$
df(error) = 24 - 4 = $\underline{20}$

c. MS(factor) = $\dfrac{4.83\overline{3}}{3} = \underline{1.61\overline{1}}$

MS(error) = $\dfrac{167.6\overline{6}}{20} = \underline{8.38\overline{3}}$

d. $F^* = \dfrac{1.61\overline{1}}{8.38\overline{3}} = \underline{0.192}$

5. a. n = 12 c = 3 k = 4
T = 42 $\Sigma(x^2)$ = 162

SS(total) = $162 - \dfrac{(42)^2}{12}$ = 162 - 147 = $\underline{15.0}$

SOLUTIONS: Chapter 12

$$\text{SS(factor)} = \left[\frac{14^2}{4} + \frac{14^2}{4} + \frac{14^2}{4}\right] - 147 = 147 - 147 = \underline{0.0}$$

$$\text{SS(error)} = 162 - 147 = \underline{15.0}$$

b. $n = 12 \qquad c = 3 \qquad k = 4$

$T = 40 \qquad \Sigma(x^2) = 152$

$$\text{SS(total)} = 152 - \frac{(40)^2}{12} = 152 - 133.\overline{3} = \underline{18.6\overline{6}}$$

$$\text{SS(factor)} = \left[\frac{12^2}{4} + \frac{20^2}{4} + \frac{8^2}{4}\right] - 133.\overline{3} = 152 - 133.\overline{3} = \underline{18.6\overline{6}}$$

$$\text{SS(error)} = 152 - 152 = \underline{0.0}$$

c. Yes. In both cases the sum of squares for the total is partitioned into its own value and zero accordingly.

d. $n = 12 \qquad c = 3 \qquad k = 4$

$T = 42 \qquad \Sigma(x^2) = 170$

$$\text{SS(total)} = 170 - \frac{(42)^2}{12} = 170 - 147 = \underline{23.0}$$

$$\text{SS(factor)} = \left[\frac{14^2}{4} + \frac{10^2}{4} + \frac{18^2}{4}\right] - 147 = 155 - 147 = \underline{8.0}$$

$$\text{SS(error)} = 170 - 155 = \underline{15.0}$$

e. "The values might double," is the most probable answer.

f. $n = 24 \qquad c = 3 \qquad k = 8$

$T = 84 \qquad \Sigma(x^2) = 340$

$$\text{SS(total)} = 340 - \frac{(84)^2}{24} = 340 - 294 = \underline{46.0}$$

$$\text{SS(factor)} = \left[\frac{28^2}{8} + \frac{20^2}{8} + \frac{36^2}{8}\right] - 294 = 310.0 - 294.0 = \underline{16.0}$$

SS(error) = 340 - 310 = 30.0

g. $n = 24 \quad c = 6 \quad k = 4$
$T = 84 \quad \Sigma(x^2) = 340$

$$SS(total) = 340 - \frac{(84)^2}{24} = 340 - 294 = \underline{46.0}$$

$$SS(factor) = \left[\frac{14^2}{4} + \frac{14^2}{4} + \frac{10^2}{4} + \frac{10^2}{4} + \frac{18^2}{4} + \frac{18^2}{4}\right] - \frac{(84)^2}{24}$$

$$= 310 - 294 = \underline{16.0}$$

SS(error) = 340 - 310 = $\underline{30.0}$

h. Each of the sums of squares found in answers f and g is exactly double the corresponding values from d. However, the number of degrees of freedom did not double in all cases and therefore the mean squares (our estimation of variance) did not double.

6. a. $n = 12 \quad c = 3 \quad k = 4$
$C_1 = 194 \quad C_2 = 190 \quad C_3 = 198$
$T = 582 \quad \Sigma(x^2) = 28{,}250$

$$SS(total) = 28{,}250 - \frac{(582)^2}{12} = 28{,}250 - 28{,}227 = \underline{23.0}$$

$$SS(factor) = \left[\frac{194^2}{4} + \frac{190^2}{4} + \frac{198^2}{4}\right] - 28{,}227 = 28{,}235 - 28{,}227 = \underline{8.0}$$

SS(error) = 28,250 - 28,235 = $\underline{15.0}$

b. The three sums of squares are identical.

c. Each piece of data in this exercise, x, is exactly 45 larger than the corresponding piece of data, u, in Illustration C ($x = u + 45$). The constant 45 does not affect measures of variance.

7. a. The null hypothesis that is to be tested using the ANOVA method is:
H_0: The variance due to the kind of gasoline is no greater than the variance due to experimental error.

The null hypothesis above is a replacement for the claim we would like to test: "The mean number of miles per gallon is the same for both kinds of gasoline."

b. Using the idea of coding as described above in Exercise 6, each piece of data has been reduced by 17.

0	4.5
1.4	1.7
2.2	3.8
3.6	10.0	13.6

$n = 6$ $c = 2$ $k = 3$ $C_1 = 3.6$ $C_2 = 10.0$

$T = 13.6$ $\Sigma(x^2) = 44.38$

$$SS(\text{total}) = 44.38 - \frac{(13.6)^2}{6} = 44.38 - 30.82\bar{6} = 13.55\bar{3}$$

$$SS(\text{factor}) = \left[\frac{3.6^2}{3} + \frac{10.0^2}{3}\right] - 30.82\bar{6} = 37.65\bar{3} - 30.82\bar{6} = 6.82\bar{6}$$

$SS(\text{error}) = 44.38 - 37.65\bar{3} = 6.72\bar{6}$

ANOVA Table:

	SS	df	MS
Factor	6.82\bar{6}	1	6.82\bar{6}
Error	6.72\bar{6}	4	1.681\bar{6}
Total	13.55\bar{3}	5	

$$F^* = \frac{6.82\bar{6}}{1.681\bar{6}} = 4.059$$

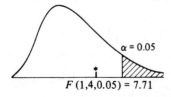

$F(1,4,0.05) = 7.71$

Decision: Fail to reject H_0.

SOLUTIONS: Chapter 12

Conclusion: The set of data does not present significant evidence that the mileage of high-test gasoline is better than the mileage of regular gasoline.

8. a. $H_0: \mu_A = \mu_B = \mu_C$ or $(\sigma_{treatment})^2/(\sigma_{error})^2 \leq 1$

b. $n = 15$ $c = 3$ $k = 5$

$C_1 = 55$ $C_2 = 33$ $C_3 = 28$

$T = 116$ $\Sigma(x^2) = 1040.0$

$$SS(total) = 1040 - \frac{(116)^2}{15} = 1040 - 897.0\bar{6} = \underline{142.9\bar{3}}$$

$$SS(factor) = \left[\frac{55^2}{5} + \frac{33^2}{5} + \frac{28^2}{5}\right] - 897.0\bar{6} = 979.6 - 897.0\bar{6} = \underline{82.5\bar{3}}$$

$SS(error) = 1040 - 979.6 = \underline{60.4}$

ANOVA Table:

	SS	df	MS
Factor	82.5$\bar{3}$	2	41.2$\bar{6}$
Error	60.4	12	5.0$\bar{3}$
Total	142.9$\bar{3}$	14	

$$F^* = \frac{41.2\bar{6}}{5.0\bar{3}} = 8.199$$

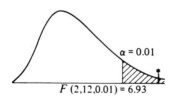

$\alpha = 0.01$

$F(2,12,0.01) = 6.93$

Decision: Reject H_0.

Conclusion: The set of data suggests that the treatment does have a significant effect on the outcome.

9. a. $H_0: \mu_A = \mu_B = \mu_C = \mu_D$

SOLUTIONS: Chapter 12

b. The set of data was coded by subtracting 70 from each entry.

	0	18	-10	1
	2	19	3	8
	9	21	9	-3
	11	10	-11	-8
	-1	24	4	8
	5		-5	
Totals	26	92	-10	6	114

$n = 22 \qquad c = 4$
$k_1 = k_3 = 6 \qquad k_2 = k_4 = 5$
$T = 114 \qquad \Sigma(x^2) = 288$

$$SS(total) = 2588 - \frac{(114)^2}{22} = 2588 - 590.72\overline{72} = \underline{1997.27\overline{27}}$$

$$SS(factor) = \left[\frac{26^2}{6} + \frac{92^2}{5} + \frac{(-10)^2}{6} + \frac{6^2}{5}\right] - 590.72\overline{72}$$

$$= 1829.33\overline{33} - 590.72\overline{72}$$
$$= \underline{1238.6060}$$

$$SS(error) = 2588 - 1829.33\overline{33} = \underline{758.66\overline{66}}$$

ANOVA Table:

	SS	df	MS
Factor	1238.61	3	412.87
Error	758.67	18	42.15
Total	1997.27	21	

$$F^* = \frac{412.87}{42.15} = 9.79$$

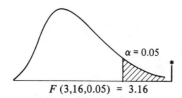

$F(3, 16, 0.05) = 3.16$

Decision: Reject H_0.
Conclusion: The set of data suggests that all four classes did not achieve at the same level.

SOLUTIONS: Chapter 12

10. a. $x_{cr} = \mu + \text{Track}_c + \epsilon_{cr}$

b. $H_0: \mu_C = \mu_I = \mu_{II}$

$n = 15 \qquad c = 3 \qquad k = 5$

$C_1 = 38.9 \qquad C_2 = 32.2 \qquad C_3 = 32.2$

$T = 103.3 \qquad \Sigma(x^2) = 724.59$

$$\text{SS(total)} = 724.59 - \frac{(103.3)^2}{15} = 724.59 - 711.39 = \underline{13.20}$$

$$\text{SS(factor)} = \left[\frac{38.9^2}{5} + \frac{32.2^2}{5} + \frac{32.2^2}{5} \right] - 711.39 = 724.59 - 711.39 = \underline{5.99}$$

SS(error) = 724.59 - 717.38 = $\underline{7.21}$

ANOVA Table:

	SS	df	MS
Factor	5.99	2	2.995
Error	7.21	12	0.6008
Total	13.20	14	

$$F^* = \frac{2.995}{0.6008} = 4.98$$

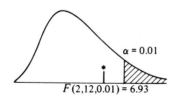

$\alpha = 0.01$

$F(2, 12, 0.01) = 6.93$

Decision: Fail to reject H_0.

Conclusion: The set of data suggests that the difference in tracks does not have a significant effect on the times.

CHAPTER 13

SOLUTIONS TO EXERCISES 13-1

1. a. $\hat{y} = 1.0 + 0.5(4) = 3.0$
   $e = y - \hat{y} = 5 - 3 = \underline{2.0}$ (above the line of best fit)

   b. $\hat{y} = 1.0 + 0.5(6) = 4.0$
   $e = y - \hat{y} = 3 - 4 = \underline{-1.0}$ (below the line of best fit)

   c. $\hat{y} = 1.0 + 0.5(8) = 5.0$
   $e = y - \hat{y} = 5 - 5 = \underline{0.0}$ (on the line of best fit)

SOLUTIONS TO EXERCISES 13-2

1. a.

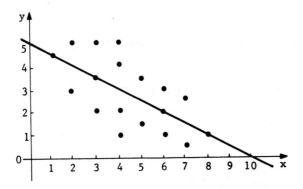

$\Sigma x = 80, \Sigma y = 50, \Sigma x^2 = 420, \Sigma xy = 190, \Sigma y^2 = 177.50$

b. $b_0 = 5.0, b_1 = -0.5; \hat{y} = 5.0 - 0.5x$

c. (1) By formula (13-6): SSE = $(0)^2 + (+1)^2 + (-1)^2 + (+1.5)^2 + (0)^2$
   $+ (-1.5)^2 + (+2)^2 + (+1)^2 + (-1)^2 + (-2)^2 + (+1)^2 + (-1)^2 + (+1)^2$
   $+ (0)^2 + (-1)^2 + (+1)^2 + (-1)^2 + (0)^2 = \underline{22.5}$

   (2) By formula (13-8): SSE = $(177.5) - (5.0)(50) - (-0.5)(190) = \underline{22.5}$

$$s^2 = \frac{SSE}{df} = \frac{22.5}{16} = \underline{1.406}$$

SOLUTIONS: Chapter 13

SOLUTIONS TO EXERCISES 13-3

1. At $x = 7$, $\hat{y} = 1.0 + 0.5(7) = 4.5$.

   a. Using formula (13-15) and information in the illustration:

   $$4.5 \pm (1.76)(1.363)\sqrt{0.0625 + 0}$$

   $$4.5 \pm (1.76)(1.363)(0.250)$$

   $$4.5 \pm 0.60$$

   (3.90 to 5.10), the 0.90 confidence estimate for $y_{x=7}$

   b. Using formula (13-16):

   $$4.5 \pm (1.76)(1.363)\sqrt{1 + 0.0625}$$

   $$4.5 \pm (1.76)(1.363)(1.03)$$

   $$4.5 \pm 2.47$$

   (2.03 to 6.97), the 0.90 confidence estimate for $y_{x=7}$

2. Information from Exercise 1 in Lesson 13-2:

   $\hat{y} = 5.0 - 0.5x$

   $n = 18$, $\Sigma x = 80$, $\Sigma x^2 = 420$, $s^2 = 1.406$

   $\bar{x} = \dfrac{80}{18} = 4.44\overline{4}$

   a. At $x = 3$, $\hat{y} = 5.0 - 0.5(3) = 3.5$.
   Using formula (13-15):

   $$3.5 \pm t(16, 0.025) \cdot \sqrt{1.406} \cdot \sqrt{\dfrac{1}{18} + \dfrac{18(3 - 4.4444)^2}{18(420) - (80)(80)}}$$

   $$3.5 \pm (2.12)(1.185)\sqrt{0.0556 + 0.0324}$$

   $$3.5 \pm (2.12)(1.185)\sqrt{0.0880}$$

   $$3.5 \pm 0.75$$

   (2.75 to 4.25), the 0.95 confidence estimate for $\mu_{y|x=3}$

b. At $x = 7, \hat{y} = 5.0 - 0.5(7) = 1.5$.
Using formula (13-15):

$$1.5 \pm (2.12)(1.185)\sqrt{0.0556 + \frac{18(7 - 4.4444)^2}{1160}}$$

$$1.5 \pm (2.12)(1.185)\sqrt{0.0556 + 0.1014}$$

$$1.5 \pm 0.9954$$

$$1.5 \pm 1.00$$

(0.50 to 2.50), the 0.95 confidence estimate for $\mu_{y|x = 3}$

c. Using formula (13-16) and the results found in part a:

$$3.5 \pm (2.12)(1.185)\sqrt{1.0880}$$

$$3.5 \pm 2.62$$

(0.88 to 6.12), the 0.95 confidence estimate for $y_{x = 3}$

d. Using formula (13-16) and the results found in part b:

$$1.5 \pm (2.12)(1.185)\sqrt{1.1570}$$

$$1.5 \pm 2.70$$

(-1.2 to 4.2), the 0.95 confidence estimate for $y_{x = 7}$

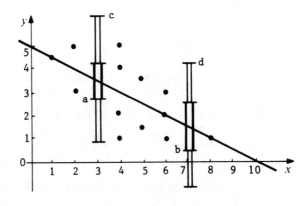

SOLUTIONS TO SELF-CORRECTING EXERCISES FOR CHAPTER 13

1. a.

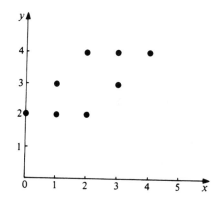

b. mean of x: $\bar{x} = \dfrac{\Sigma x}{n} = \dfrac{16}{8} = \underline{2.0}$

mean of y: $\bar{y} = \dfrac{\Sigma x}{n} = \dfrac{24}{8} = \underline{3.0}$

c.

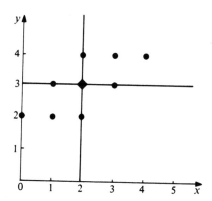

d. The covariance is calculated by use of formula (13-1) and with the aid of the following table.

SOLUTIONS: Chapter 13

x	y	$x - \bar{x}$	$y - \bar{y}$	$(x - \bar{x})(y - \bar{y})$
0	2	-2	-1	2
1	2	-1	-1	1
1	3	-1	0	0
2	2	0	-1	0
2	4	0	1	0
3	3	1	0	0
3	4	1	1	1
4	4	2	1	2
16	24	0 (ck)	0 (ck)	6

$$\text{covar}(x, y) = \frac{\Sigma[(x - \bar{x})(y - \bar{y})]}{n - 1} = \frac{6}{7} = \underline{0.857}$$

2. a.

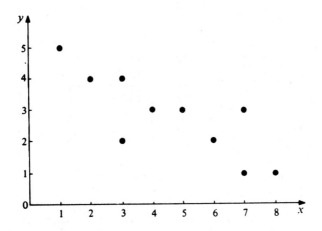

b. mean of $x = \bar{x} = \dfrac{\Sigma x}{n} = \dfrac{46}{10} = \underline{4.6}$

mean of $y = \bar{y} = \dfrac{\Sigma y}{n} = \dfrac{28}{10} = \underline{2.8}$

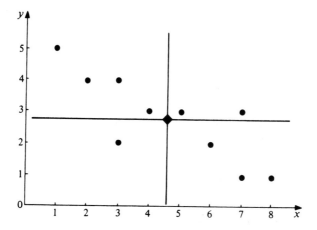

c. The covariance is calculated by use of formula (13-1) and with the aid of the following table.

x	y	$x - \bar{x}$	$y - \bar{y}$	$(x - \bar{x})(y - \bar{y})$
1	5	-3.6	2.2	-7.92
2	4	-2.6	1.2	-3.12
3	2	-1.6	-0.8	1.28
3	4	-1.6	1.2	-1.92
4	3	-0.6	0.2	-0.12
5	3	0.4	0.2	0.08
6	2	1.4	-0.8	-1.12
7	1	2.4	-1.8	-4.32
7	3	2.4	0.2	0.48
8	1	3.4	-1.8	-6.12
46	28	0	0	-22.80

$$\text{covar}(x, y) = \frac{-22.80}{9} = -2.53\overline{3} = \underline{-2.53}$$

3. The calculations below require the totals obtained from the extensions table; therefore, we might as well complete that table now so that we will have the required totals.

x	y	x^2	xy	y^2
0	2	0	0	4
1	2	1	2	4
1	3	1	3	9
2	2	4	4	4
2	4	4	8	16
3	3	9	9	9
3	4	9	12	16
4	4	16	16	16
16	24	44	54	78

a. The calculation of the standard deviation of x, s_x, and the standard deviation of y, s_y, is accomplished by using formula (2-10) and then taking the square root.

$$s_x = \sqrt{\frac{44 - \frac{(16)(16)}{8}}{(7)}} = \sqrt{\frac{12}{7}} = \sqrt{1.714} = \underline{1.309}$$

$$s_y = \sqrt{\frac{78 - \frac{(24)(24)}{8}}{(7)}} = \sqrt{\frac{6}{7}} = \sqrt{0.857} = \underline{0.926}$$

b. $r = \dfrac{\text{covar}(x,y)}{s_x \cdot s_y} = \dfrac{0.857}{(1.309)(0.926)} = \underline{0.707}$

c. $SS(xy) = 54 - \dfrac{(16)(24)}{8} = 54 - 48 = 6.0$

$SS(x) = 12$, $SS(y) = 6$ from part (b)

$r = \dfrac{6}{\sqrt{(12)(6)}} = \underline{0.707}$

d. The answers found in parts b and c are the same, as they should be. Formula (13-3) is a rewriting of the definition formula and is equivalent to formula (13-2).

SOLUTIONS: Chapter 13

4. The questions below require the five totals obtained from the extensions table showing x^2, xy, and y^2; therefore, we will find these totals first.

x	y	x^2	xy	y^2
1	5	1	5	25
2	4	4	8	16
3	2	9	6	4
3	4	9	12	16
4	3	16	12	9
5	3	25	15	9
6	2	36	12	4
7	1	49	7	1
7	3	49	21	9
8	1	64	8	1
46	28	262	106	94

a. The calculation of the standard deviation of x, s_x, and the standard deviation of y, s_y, is accomplished by using formula (2-10) and then taking the square root.

$$s_x = \sqrt{\frac{(262) - \frac{(46)(46)}{10}}{(9)}} = \sqrt{\frac{50.4}{9}} = \sqrt{5.60} = \underline{2.366}$$

$$s_y = \sqrt{\frac{(94) - \frac{(28)(28)}{10}}{(9)}} = \sqrt{\frac{15.6}{9}} = \sqrt{1.733} = \underline{1.316}$$

b. $r = \dfrac{\text{covar}(x, y)}{s_x \cdot s_y} = \dfrac{-2.53}{(2.366)(1.316)} = \underline{-0.813}$

c. $SS(xy) = 106 - \dfrac{(46)(28)}{10} = 106 - 128.8 = -22.8$

$r = \dfrac{-22.8}{\sqrt{(50.4)(15.6)}} = \underline{-0.813}$

d. Yes, the answers obtained in parts b and c are the same. Formula (13-3) is an equivalent rewriting of the definition formula (13-2); therefore, they must yield like answers.

5. a. $H_0: \rho = 0$
   $H_a: \rho > 0$

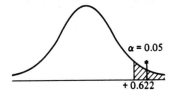

The calculated value of r, r^*, obtained in Exercise 3, was 0.707. Since +0.707 falls in the critical region, we reach a decision of reject H_0. This means we conclude that the observed set of data shows a significant amount of linear correlation.

b. To determine the 95 percent confidence-interval estimate for ρ, we use Table 10 as described on page 443 in the textbook. We enter the table at +0.7 along the bottom of the graph and follow that line up until we cross the two belts marked with the number 8. We read across to the side of the table and find the values −0.02 and +0.92. Thus the 95 percent confidence interval for ρ is

(−0.02 to 0.92)

Note: This interval contains the value zero, which would generally mean that the corresponding null hypothesis should not have been rejected. This situation occurs as a result of a one-tailed hypothesis test and a two-tailed confidence interval.

6. a. $H_0: \rho = 0$
   $H_a: \rho \neq 0$

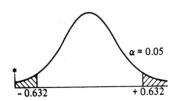

$r^* = -0.813$ (as calculated in Exercise 4)
Decision: Reject H_0.
Conclusion: There is evidence of linear correlation in this set of data.

b. (See solution to Exercise 5b for instruction on using Table 10.)

(−0.95 to −0.33) 95 percent confidence interval for ρ

SOLUTIONS: Chapter 13

7. a. The summations necessary for using formulas (3-6) and (3-7) are found in the answer to Exercise 3 above.
$n = 8, \Sigma x = 16, \Sigma y = 24, \Sigma x^2 = 44, \Sigma xy = 54, \Sigma y^2 = 78$

$$b_1 = \frac{SS(xy)}{SS(x)} = \frac{6.0}{12.0} = 0.5$$

$$b_0 = \frac{1}{8}[24 - (0.5)(16)] = 2.0$$

$$\hat{y} = 2.0 + 0.5x$$

Recall that the equation of a line is the algebraic expression that tells the relationship between the abscissa (x-value) and the ordinate ($\hat{y}$-value) of every point on that line. Therefore, to find the value of $\hat{y}$ that corresponds to each specific x-value, you need only to replace the x with the particular number value of x, and the resulting value will be that of $\hat{y}$.

b. $\hat{y} = 2.0 + 0.5x$; this says that the value of $\hat{y}$ is 2.0 more than one-half the value of x.

If $x = 0$, then $\hat{y} = 2.0 + 0.5(0) = \underline{2.0}$
If $x = 1$, then $\hat{y} = 2.0 + 0.5(1) = \underline{2.5}$
If $x = 2$, then $\hat{y} = 2.0 + 0.5(2) = \underline{3.0}$
If $x = 3$, then $\hat{y} = 2.0 + 0.5(3) = \underline{3.5}$
If $x = 4$, then $\hat{y} = 2.0 + 0.5(4) = \underline{4.0}$

c.

x	y	$\hat{y}$	$e = y - \hat{y}$
0	2	2.0	0.0
1	2	2.5	-0.5
1	3	2.5	0.5
2	2	3.0	-1.0
2	4	3.0	1.0
3	3	3.5	-0.5
3	4	3.5	0.5
4	4	4.0	0.0

The values of $\hat{y}$ on the chart above were obtained as answers in part b of this exercise.

d. $s_e^2 = \dfrac{\Sigma(y - \hat{y})^2}{n - 2}$

$= \dfrac{(0)^2 + (-0.5)^2 + (0.5)^2 + (-1.0)^2 + (1.0)^2 + (-0.5)^2 + (0.5)^2 + (0)^2}{8 - 2}$

$$s_e^2 = \frac{3}{6} = \underline{0.5}$$

The value of $\Sigma(y - \hat{y})^2$ is found by squaring each e on the chart shown in the answer to part c and then summing these squared quantities.

e. $s_e^2 = \dfrac{\Sigma y^2 - b_0(\Sigma y) - b_1(\Sigma xy)}{n - 2} = \dfrac{78 - (2.0)(24) - (0.5)(54)}{8 - 2}$

$= \dfrac{78 - 48 - 27}{6} = \dfrac{3}{6}$

$= \underline{0.50}$

f. The answers in part b and c are identical. This will always happen provided the values used for b_0 and b_1 are the exact values. If you use a number which has been rounded off, then this round-off error will be magnified by the products taken and formula (13-8) will be inaccurate. To avoid this, you should do the calculations using several extra decimal places of accuracy.

8. a. The summations to be used with formulas (3-6) and (3-7) are found in the answer to Exercise 4 above.
$n = 10$, $\Sigma x = 46$, $\Sigma y = 28$, $\Sigma x^2 = 262$, $\Sigma xy = 106$, $\Sigma y^2 = 94$

$b_1 = \dfrac{SS(xy)}{SS(x)} = \dfrac{-22.8}{50.4} = -0.452 = -0.45$

$b_0 = \dfrac{1}{10}[28 - (-0.452)(46)] = 4.879 = 4.88$

$\hat{y} = 4.88 - 0.45x$

(See Exercise 7a.)

b. If $x = 1$, then $\hat{y} = 4.88 - 0.45(1) = \underline{4.43}$
If $x = 2$, then $\hat{y} = 4.88 - 0.45(2) = \underline{3.98}$
If $x = 3$, then $\hat{y} = 4.88 - 0.45(3) = \underline{3.53}$
If $x = 4$, then $\hat{y} = 4.88 - 0.45(4) = \underline{3.08}$
If $x = 5$, then $\hat{y} = 4.88 - 0.45(5) = \underline{2.63}$
If $x = 6$, then $\hat{y} = 4.88 - 0.45(6) = \underline{2.18}$
If $x = 7$, then $\hat{y} = 4.88 - 0.45(7) = \underline{1.73}$
If $x = 8$, then $\hat{y} = 4.88 - 0.45(8) = \underline{1.28}$

SOLUTIONS: Chapter 13

c.

x	y	$\hat{y}$	$e = y - \hat{y}$
1	5	4.43	0.57
2	4	3.98	0.02
3	2	3.53	-1.53
3	4	3.53	0.47
4	3	3.08	-0.08
5	3	2.63	0.37
6	2	2.18	-0.18
7	1	1.73	-0.73
7	3	1.73	1.27
8	1	1.28	-0.28

d. Using formula (13-6):

$$s_e^2 = \frac{(0.57)^2 + (0.02)^2 + (1.53)^2 + \ldots + (-0.28)^2}{10 - 2} = \frac{5.287}{8}$$

$$s_e^2 = 0.660875 = \underline{0.66}$$

Using formula (13-8):

$$s_e^2 = \frac{94 - (4.88095)(28) - (-0.45238)(106)}{10 - 2} = \frac{5.28568}{8}$$

$$s_e^2 = 0.66071 = \underline{0.66}$$

e. The results in part b are the same. The value of s_e^2 is the same, except for the round-off error, when calculated by either equation.

9. a. The standard error of b_1 is calculated using the square root of formula (13-11)

$$s_{b_1}^2 = \frac{s_e^2}{SS(x)} = \frac{(0.50)}{12} = 0.0416\overline{6}$$

Note: s_e^2 was calculated in Exercise 7d.

$$s_{b_1} = \sqrt{s_{b_1}^2} = \sqrt{0.0416\overline{6}} = \underline{0.204}$$

b. $H_0: \beta_1 = 0$
$H_a: \beta_1 > 0$

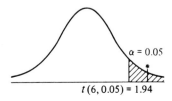

$\alpha = 0.05$

$t(6, 0.05) = 1.94$

$$t^* = \frac{b_1 - \beta_1}{s_{b_1}} = \frac{0.500}{0.204}$$

$$t^* = \underline{2.45}$$

Decision: Reject H_0.
Conclusion: The observed value of the slope is significantly greater than zero.

c. $b_1 \pm t(n-2, \alpha/2) \cdot s_{b_1}$ [formula (13-13)]

$0.50 \pm (2.45)(0.204)$

0.50 ± 0.500

($\underline{0.00}$ to $\underline{1.00}$), the 0.95 confidence interval for β_1

d.

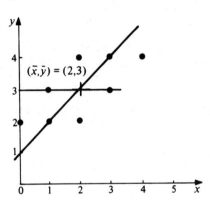

10. a. See comments in Exercise 9.

$$s_{b_1} = \sqrt{\frac{(0.661)}{50.4}} = \sqrt{0.013115}$$

$s_{b_1} = \underline{0.1145}$

b. $H_0: \beta_1 = 0$
$H_a: \beta_1 \neq 0$
$\alpha = 0.05$

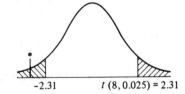

$t^* = \dfrac{-0.45}{0.1145} = \underline{-3.93}$

Decision: Reject H_0.
Conclusion: There is a significant amount of slope observed in this sample.

c. $-0.45 \pm (2.31)(0.1145)$
-0.45 ± 0.26
($\underline{-0.71}$ to $\underline{-0.19}$) for β_1

d.

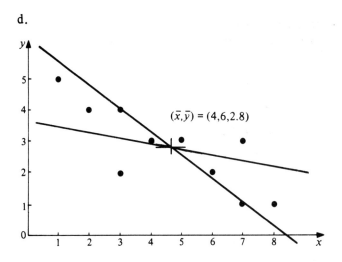

11. a. In order to use formula (13-15) to calculate the confidence-interval estimate, we need several values previously calculated.

Equation of regression line: $\hat{y} = 2.0 + 0.5x$
Standard deviation: $s = 0.707$
$n = 8$
$\bar{x} = 2.0$
$SS(x) = 12$

At $x = 1.0$ $(x_0 = 1)$, $\hat{y} = 2.0 + 0.5(1.0) = 2.5$. This value is the point estimate for $\mu_{y|x = 1}$ and therefore becomes the center point of the interval estimate.

$$\hat{y} \pm t(n-2, \alpha/2) \cdot s \cdot \sqrt{\frac{1}{n} + \frac{(x_0 - \bar{x})^2}{12}}$$

$$2.50 \pm (2.45)(0.707) \sqrt{\frac{1}{8} + \frac{(1-2)^2}{12}}$$

$2.50 \pm (2.45)(0.707) \sqrt{0.12500 + 0.08333}$

$2.50 \pm (2.45)(0.707) \sqrt{0.2083}$

$2.50 \pm (2.45)(0.707)(0.456)$

2.50 ± 0.79

(1.71 to 3.29) for $\mu_{y|x = 1}$

b. To calculate other confidence intervals, we need to obtain a different point estimate, $\hat{y}$ (these may be found in the answer to Exercise 7c), and to change the value of x_0.

At $x = 0, \hat{y} = 2.0$.
Interval estimate:

$$2.0 \pm (2.45)(0.707)\sqrt{0.12500 + \frac{(0-2)^2}{12}}$$

$$2.0 \pm (2.45)(0.707)\sqrt{0.4583}$$

$$2.0 \pm (2.45)(0.707)(0.677)$$

$$2.0 \pm 1.17$$

(0.83 to 3.17) for $\mu_{y|x} = 0$

At $x = 2, \hat{y} = 3.0$.
Interval estimate:

$$3.0 \pm (2.45)(0.707)\sqrt{0.1250 + 0}$$

$$3.0 \pm (2.45)(0.707)(0.354)$$

$$3.0 \pm 0.61$$

(2.39 to 3.61) for $\mu_{y|x} = 2$

At $x = 3, \hat{y} = 3.5$.

$$3.50 \pm (2.45)(0.707)\sqrt{0.12500 + \frac{(3-2)^2}{12}}$$

$$3.50 \pm (2.45)(0.707)\sqrt{0.12500 + 0.08333}\ \}\qquad \text{(same product as estimated at } x = 1.0\text{)}$$

$$3.50 \pm 0.79$$

(2.71 to 4.29) for $\mu_{y|x} = 3$

At $x = 4, \hat{y} = 4.0$.

$$4.00 \pm (2.45)(0.707)\sqrt{0.12500 + \frac{(4-2)^2}{12}}$$

$$4.00 \pm (2.45)(0.707)\sqrt{0.12500 + 0.33333}$$

$$4.00 \pm 1.17$$

(2.83 to 5.17) for $\mu_{y|x} = 4$

c.

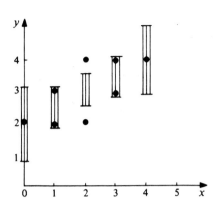

d.

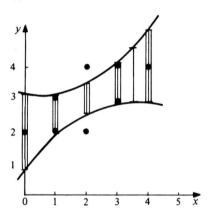

12. See comments in answer to Exercise 11. Use formula (13-16).
    a. $\hat{y} = 4.88 - 0.45$; $s = \sqrt{0.66} = 0.8124$
    $n = 10$; $\bar{x} = 4.6$; $n(\Sigma x^2) - (\Sigma x)^2 = 504$

    At $x = 3$, $y = 3.53$.
    Interval estimate:

    $$3.53 \pm (2.31)(0.8124) \sqrt{1 + \frac{1}{10} + \frac{(3 - 4.6)^2}{50.4}}$$

    $3.53 \pm (2.31)(0.8124) \sqrt{1 + 0.10000 + 0.05079}$

    $3.53 \pm (2.31)(0.8124)(1.0727)$

    3.53 ± 2.01

    $\underline{(1.52 \text{ to } 5.54)}$ for $y_{x=3}$

SOLUTIONS: Chapter 13

b. At $x = 1, \hat{y} = 4.43$.

$$4.43 \pm (2.31)(0.8124)\sqrt{1 + 0.10000 + \frac{(1 - 4.6)^2}{50.4}}$$

$$4.43 \pm (2.31)(0.8124)\sqrt{1 + 0.10000 + 0.25714}$$

$$4.43 \pm 2.19$$

(2.24 to 6.62) for $y_{x=1}$

At $x = 5, \hat{y} = 2.63$.
Interval:

$$2.63 \pm (2.31)(0.8124)\sqrt{1 + 0.1000 + \frac{(5 - 4.6)^2}{50.4}}$$

$$2.63 \pm (2.31)(0.8124)(1.0503)$$

$$2.63 \pm 1.97$$

(0.66 to 4.60) for $y_{x=5}$

At $x = 7, \hat{y} = 1.73$.
Interval:

$$1.73 \pm (2.31)(0.8124)\sqrt{1 + 0.100 + \frac{(7 - 4.6)^2}{50.4}}$$

$$1.73 \pm (2.31)(0.8124)(1.1019)$$

$$1.73 \pm 2.07$$

(−0.34 to 3.80) for $y_{x=7}$

c.

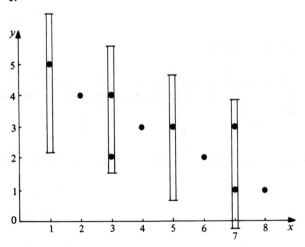

d.

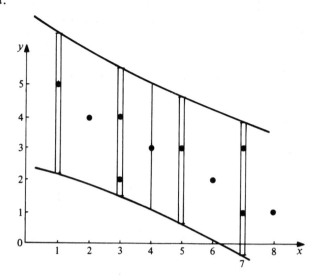

(1.1 to 5.0), 0.95 confidence interval for $y_{x=4}$

CHAPTER 14

SOLUTIONS TO EXERCISES 14-1

1. If $x = 4$ becomes part of the critical region, then the value of α actually becomes 0.118 and exceeds the 0.10 as specified. Thus, the critical value is 3, just as with $\alpha = 0.05$.

2. The addition rule for the probability of mutually exclusive events.

3. a. $x = 0, 1, 2, 3, \ldots, 9$
   b. $x = 0, 1, 2, \ldots, 9$, and $21, 22, 23, \ldots, 30$. (both tails)
      $x = 22$ means that the number of the least-frequent sign is only 8.

4. $n = 50$

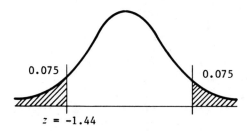
$z = -1.44$

$$z = \frac{x' - n/2}{\sqrt{n}/2}$$

$$-1.44 = \frac{x' - 25}{\sqrt{50}/2}$$

$$x' - 25 = \left(\frac{7.07}{2}\right)(-1.44)$$

$x'\ \ \ = 25 - 5.09$

$x'\ \ \ = 19.9$

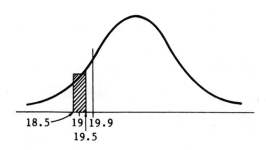
18.5 19 19.9
      19.5

Critical value: $x = 19$. (The critical region: $x \leqslant 19$.)

SOLUTIONS TO EXERCISES 14-2

1.

	Miles per gallon	Rank	R_a, regular	R_b, supreme
1	21	1.5	1.5	
2	21	1.5		1.5
3	25	3.0	3.0	
4	26	4.5	4.5	
5	26	4.5		4.5
6	29	6.0		6.0
7	30	7.5	7.5	
8	30	7.5		7.5
9	31	9.0		9.0
10	32	10.0		10.0
11	33	11.0		11.0
12	36	12.0	12.0	
13	37	13.5	13.5	
14	37	13.5		13.5
15	40	15.0		15.0
16	41	16.5	16.5	
17	41	16.5	16.5	
18	43	18.5	18.5	
19	43	18.5		18.5
20	44	20.0	20.0	
21	45	21.0	21.0	
22	47	22.0		22.0
Sum			134.5	118.5
			253.0	

H_0: Gas mileage is the same from both grades (no difference).
H_a: Gas mileage is not the same.

From Table 10 we find the critical value of $U = 30$.

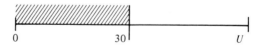

Now calculate U:

$$U_a = n_a \cdot n_b + \frac{(n_b)(n_b + 1)}{2} - R_b$$

$$U_b = n_a \cdot n_b + \frac{(n_a)(n_a + 1)}{2} - R_a$$

$$U_a = 11 \cdot 11 + \frac{(11)(12)}{2} - 118.5 = 121 + \frac{132}{2} - 118.5$$

SOLUTIONS: Chapter 14

$$= 121 + 66 - 118.5 = 68.5$$

$$U_b = 11 \cdot 11 + \frac{(11)(12)}{2} - 134.5 = 121 + 66 - 134.5 = 52.5$$

Therefore, $U = \underline{52.5}$.
Decision: Fail to reject H_0.
Conclusion: There is not sufficient evidence to reject the null hypothesis that there is no difference.

SOLUTIONS TO EXERCISES 14-3

1. H_0: Dining sites are random.
   H_a: Dining sites are not random.
   $n_F = 21$, $n_H = 9$, and $n_F > 20$; therefore, we use z.

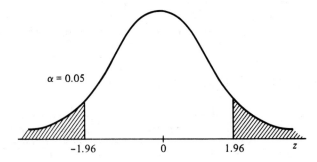

The runs are:

FFF H F H F H FFF H FFFFF H F HHH FFFFF H FF

Therefore, $V = 15$.
Now we use formulas (14-8), (14-9), and (14-10).

$$\mu_V = \frac{2n_F n_H}{n_F + n_H} + 1 = \frac{(2)(21)(9)}{21 + 9} + 1 = \frac{378}{30} + 1 = 12.6 + 1 = 13.6$$

$$\sigma_V = \sqrt{\frac{2n_F n_H (2n_F n_H - n_F - n_H)}{(n_F + n_H)^2 (n_F + n_H - 1)}} = \sqrt{\frac{(378)(378 - 21 - 9)}{(900)(29)}}$$

$$= \sqrt{\frac{131,544}{26,100}}$$

$$= \sqrt{5.044} = 2.245$$

$$z = \frac{V - \mu_V}{\sigma_V} = \frac{15 - 13.6}{2.245} = \underline{0.624}$$

Decision: Fail to reject H_0:
Conclusion: Dining site selections are random.

2. H_0: The sequence of calls received is random with respect to man or woman caller.
H_a: The sequence of calls is not random.
$\alpha = 0.05$, $n_M = 5$, and $n_W = 4$; therefore we use V as a test statistic and Table 13 of Appendix D to obtain critical values.

The runs are:

WWW MM W MMM

There are four runs, so $V = \underline{4}$.
Decision: Fail to reject H_0.
Conclusion: There is no evidence to reject the null hypothesis of randomness.

SOLUTIONS TO EXERCISES 14-4

1.

Player	x, rank from fans	y, rank from media	d	d²
A	1	1	0	0
B	2	3	-1	1
C	3	2	1	1
D	4	5	-1	1
E	5	4	1	1
Sum				4

H_0: Fan rankings and media rankings are independent.
H_a: There is a positive correlation.

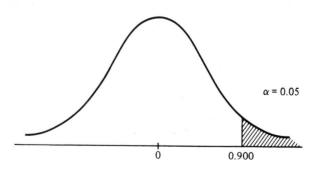

$$r_s = 1 - \frac{6(4)}{5(5^2-1)} = \frac{1-24}{120} = 1 - 0.2 = \underline{0.8}$$

Decision: Fail to reject H_0.
Conclusion: There is no evidence of positive correlation between rankings.

2. $H_0: \rho = 0$
   $H_a: \rho \neq 0$

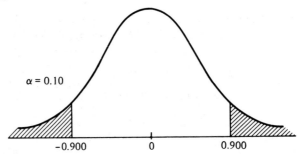

Player	Salary rank	Media rank	d^2
A	1	1	0
B	3	3	0
C	2	2	0
D	5	5	0
E	4	4	0
Sum			0

$$r_s^* = 1 - \frac{6(0)}{5(5^2-1)} = \underline{1}$$

Decision: Reject H_0.
Conclusion: There is sufficient evidence to indicate a correlation between salary rankings and media rankings.

SOLUTIONS TO SELF-CORRECTING EXERCISES FOR CHAPTER 14

1. A null hypothesis like $\mu = 100$ is tested using the sign test by assigning a sign (+ or -) to each piece of data in the sample. If the value of the piece of data is greater than 100, it is assigned a +; if it is smaller than 100, it is assigned a -. If it happens that a piece of data is exactly equal to 100, then we will assign a zero. The test statistic is the number of the least frequent sign (+ or -).

The number of signs and the kind of sign that predominates dictate the decision to be reached about the null hypothesis. If there is an overwhelming majority of pluses, then we will conclude that the mean is larger than 100. If the vast majority of the signs are negative, then we will conclude that the true value of μ is less than 100. However, if the number of plus and negative signs are approximately equal, then we will conclude that the true value of the mean is 100, since the values are approximately equally scattered around 100.

2. With the difference between two dependent means, the pieces of data from the two samples are paired; therefore, the difference between their values can be assigned a plus or minus sign depending on which value is the larger.

   With the difference between two independent means, there is no way to obtain a sign. Thus the sign test is inappropriate for the independent means tests. (Random pairing is out, as there is no guarantee that both samples are of the same size. Besides, what purpose would random pairs serve?)

3. One of the major differences between Spearman's rank correlation and Pearson's linear correlation is the fact that rank correlation only measures a "which-one-came-next" idea, while Pearson's linear correlation coefficient deals more directly with the size of the value.

4. $H_0: p = P(A) = 0.5$ (no preference).
   $H_a: p < 0.5$ (customers prefer B to A).
   $\alpha = 0.05$ and $n = 11$; from Table 10, the critical value is 2.
   The test statistic is $x = 3$.
   Decision: Fail to reject H_0.
   Conclusion: No significant preference is shown.

5. $H_0: p = P(\text{ad } 1) = 0.5$ (no preference).
   $H_a: p < 0.5$.
   $\alpha = 0.05$ and $n = 8$; Table 11 indicates the critical value is 0.
   The test statistic is $x = n(\text{prefer ad } 1)$; so $x = 2$.
   Decision: Fail to reject H_0.
   Conclusion: No preference is shown.

6. H_0: The average value is the same for both sampled populations.
   H_a: The average values are not the same.
   $\alpha = 0.10$, $n_A = 10$, and $n_B = 10$. From Table 12, page 535 of the textbook, the critical value is 27. Therefore, the critical region includes those values of $U \leqslant 27$.

Data	Initial rank	Adjusted rank	Ranks for A	Ranks for B
2	1 ⎤ $\frac{1+2}{2}$	1.5		1.5
2	2 ⎦	1.5		1.5
3	3 ⎤	4	4	
3	4 $\frac{3+4+5}{3}$	4	4	
3	5 ⎦	4	4	
4	6	6		6
5	7 ⎤	8	8	
5	8 $\frac{7+8+9}{3}$	8	8	
5	9 ⎦	8		8
6	10 ⎤ $\frac{10+11}{2}$	10.5	10.5	
6	11 ⎦	10.5		10.5
7	12 ⎤ $\frac{12+13}{2}$	12.5		12.5
7	13 ⎦	12.5		12.5
8	14 ⎤	15	15	
8	15 $\frac{14+15+16}{3}$	15	15	
8	16 ⎦	15		15
9	17 ⎤	18.5	18.5	
9	18 $\frac{17+18+19+20}{4}$	18.5	18.5	
9	19	18.5		18.5
9	20 ⎦	18.5		18.5
Sum			105.5	104.5

$$U_A = (10)(10) + \frac{(10)(11)}{2} - 104.5 = 50.5$$

$$U_B = (10)(10) + \frac{(10)(11)}{2} - 105.5 = 49.5$$

SOLUTIONS: Chapter 14

Note: $U_A + U_B$ must equal $n_A \times n_B$; use this as a check.
Therefore, $U = \underline{49.5}$.
Decision: Fail to reject H_0 ($U = 49.5$, which is greater than 27; therefore, U is not in the critical region).
Conclusion: We cannot reject the null hypothesis that the populations have the same average values.

7. H_0: There is randomness in the sequence of jogging and nonjogging.
   H_a: There is a lack of randomness in the data.
   $\alpha = 0.05$, $n_J = 37$, and $n_N = 24$; therefore we must use the z-statistic.

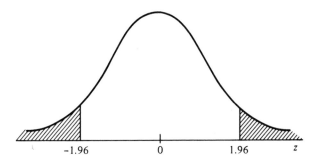

The runs are:

JJJJ N J N JJ NNNNNN JJ N JJ NN J N J NNN J N JJJJ N JJ N JJ

N J N JJ NN J N JJJJJ N JJJJJJ

Thus $V = \underline{31}$.

$$\mu_V = \frac{(2)(37)(24)}{37 + 24} + 1 = 29.11 + 1 = 30.11$$

$$\sigma_V = \sqrt{\frac{(2)(37)(24)[(2)(37)(24) - 37 - 24]}{(37 + 24)^2(37 + 24 - 1)}} = 3.693$$

$$z^* = \frac{31 - 30.11}{3.693} = \underline{0.241}$$

Decision: Fail to reject H_0.
Conclusion: There is not sufficient evidence to reject the null hypothesis of randomness.

8. a. First rank the two sets of data separately.

Math			Verbal		
Score	Initial rank	Adjusted rank	Score	Initial rank	Adjusted rank
400	1	1	350	1	1
440	2	2	360	2	2
450	3⎤	4.5	380	3	3
450	4 ⎥ $\frac{3+4+5+6}{4}$	4.5	390	4 ⎤ $\frac{4+5}{2}$	4.5
450	5 ⎥	4.5	390	5 ⎦	4.5
450	6⎦	4.5	420	6	6
460	7	7	430	7	7
470	8 ⎤ $\frac{8+9}{2}$	8.5	440	8	8
470	9 ⎦	8.5	450	9 ⎤ $\frac{9+10}{2}$	9.5
480	10	10	450	10 ⎦	9.5
560	11	11	520	11	11

Next form pairs of ranks, find d and d^2 for each pair and Σd^2.

Math, x	Verbal, y	Rank of x	Rank of y	d	d^2
450	390	4.5	4.5	0.0	0.00
440	350	2.0	1.0	1.0	1.00
460	450	7.0	9.5	-2.5	6.25
450	450	4.5	9.5	-5.0	25.00
480	380	10.0	3.0	7.0	49.00
450	520	4.5	11.0	-6.5	42.25
470	360	8.5	2.0	6.5	42.25
470	390	8.5	4.5	4.0	16.00
450	420	4.5	6.0	-1.5	2.25
560	440	11.0	8.0	3.0	9.00
400	430	1.0	7.0	-6.0	36.00
Sum					229.00

$$r_s = 1 - \frac{6(229.0)}{11(120)} = 1 - 1.041 = \underline{-0.041}$$

b. H_0: Math and verbal scores are independent.
   H_a: Math and verbal scores are not independent.
   $\alpha = 0.05$ and $n = 11$; we use Table 14 (page 537 of textbook) to find the critical values.

SOLUTIONS: Chapter 14

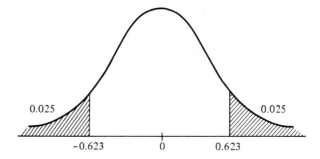

The test statistic is $r_s^* = -0.041$ (answer to part a above).
Decision: Fail to reject H_0.
Conclusion: We cannot reject the null hypothesis that math and verbal scores are independent.

9. H_0: Typing speed and accuracy are independent ($\rho = 0$).
H_a: Typing speed and accuracy are positively correlated ($\rho > 0$).

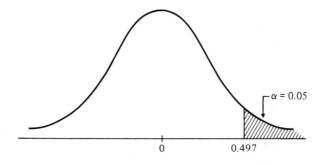

x, typing speed rank	y, accuracy rank	d	d^2
5	7	-2	4.0
6	5	1	1.0
1	2	-1	1.0
3.5	3	0.5	0.25
11	11	0	0
12	12	0	0
3.5	4	-0.5	0.25
9.5	10	-0.5	0.25
9.5	9	0.5	0.25
2	1	1.0	1.0
8	8	0	0
7	6	1.0	1.0
Sum			9.00

$$r_s^* = 1 - \frac{6(9)}{12(12^2 - 1)} = 1 - \frac{54}{1716} = \underline{0.969}$$

Decision: Reject H_0 (r_s^* is in the critical region).
Conclusion: There is sufficient evidence to indicate a correlation between typing speed and accuracy.